Programmable Controllers Using Allen-Bradley SLC 500 and ControlLogix

Programmable Controllers Using Allen-Bradley SLC 500 and ControlLogix

Robert Filer

George Leinonen

Prentice
Hall

Upper Saddle River, New Jersey
Columbus, Ohio

Library of Congress Cataloging-in-Publication Data

Filer, Robert.
    Programmable controllers using Allen-Bradley SLC 500 and ControlLogix/Robert
Filer, George Leinonen.
    p. cm.
    Includes index.
    ISBN 0-13-025603-X
    1. Programmable controllers. 2. Logic design. I. Leinonen, George. II. Title.
TJ223.P76 F543 2002
629.8'9—dc21
                                          2001021674

Editor in Chief: Stephen Helba
Assistant Vice President and Publisher: Charles E. Stewart, Jr.
Assistant Editor: Delia K. Uherec
Production Editor: Tricia L. Rawnsley
Design Coordinator: Robin G. Chukes
Cover Designer: Diane Ernsberger
Cover art: Marjory Dressler
Production Manager: Matthew Ottenweller

This book was set in Times by Carlisle Communications, Ltd. It was printed by R. R. Donnelley & Sons
Company. The cover was printed by The Lehigh Press, Inc.

Pearson Education Ltd., *London*
Pearson Education Australia Pty. Limited, *Sydney*
Pearson Education Singapore Pte. Ltd.
Pearson Education North Asia Ltd., *Hong Kong*
Pearson Education Canada, Ltd., *Toronto*
Pearson Educación de Mexico, S.A. de C.V.
Pearson Education—Japan, *Tokyo*
Pearson Education Malaysia Pte. Ltd.
Pearson Education, *Upper Saddle River, New Jersey*

10 9 8 7 6 5 4 3 2 1
ISBN: 0-13-025603-X

Dedication

To my wife, Barbara, and my children, Mark, Kristan, Anne, and Sarah.

—Robert Filer

To Siiri, Aliisa, and Stiina Mason, my granddaughters. They make me realize every day how wonderful life can be.

—George Leinonen

Preface

George and I wrote a previous book entitled *Programmable Controllers and Designing Sequential Logic.* It has been successful, but the material is now dated. We both teach programmable controller courses; George teaches at Rockwell Automation/Allen Bradley and I teach at Michigan Technological University. Our combined industrial and academic experiences have enabled us to write a book that will be useful to those in a classroom and those in the field. Since the publication of our original book, Allen Bradley merged with Rockwell Automation and introduced many new versions of equipment and software. *Programmable Controllers Using Allen-Bradley SLC 500 and ControlLogix* covers these changes in programmable controller technology and introduces automatic control.

I have used drafts of the text in my classes at Michigan Tech, where my students have been very helpful in testing the accuracy and usefulness of the material. The exercises in each chapter have been assigned to students, who have worked out all the bugs. This book is designed for engineering, engineering technology, and science students, in both electrical and mechanical programs at the associate, bachelor, master, and PhD levels. While teaching at Allen-Bradley, George has had an even broader group of students from industry.

This book is intended to be taught over the course of two college semesters, the first semester to introduce programmable controllers and the second semester for advanced programmable controllers. We have covered both sequential control and automatic control via the PID instruction. There are many example programs, most written using the SLC 500 and the ControlLogix instruction set. Upon finishing the course, students are able to write complex programming using state diagramming, zone control, subroutines, and sequential function charts.

▶ OVERVIEW

The best way to learn about programmable controllers is the hands-on approach of writing control programs. This textbook is intended to be used in this way. Chapter 2, *Interfacing and Ladder-Logic Fundamentals* and Appendix A, *Relay Control,* are designed to give students enough information to get started with some simple I/O instructions in the first week of class. I recommend 2 hours of lecture and 3 hours of laboratory time writing and executing programs each week. Chapters 3 through 9 systematically take students through SLC 500 and ControlLogix architecture and instruction sets. While covering this material, students learn specific instruction sets each week at the rate of one chapter per week. Chapter 10, *Structured Programming on the PLC®,* is designed to be used in both introductory and advanced courses. The introductory course covers state diagrams implemented via zone and subroutine. Sequential function charts are usually covered in the advanced course. Chapters 10 through 13 are used for the advanced course, in which students are assigned control problems that may each take 3 to 4 weeks to complete. Chapter 12, *Automatic Control Using PID Instruction,* introduces automatic control, scaling, and PID tuning. Upon finishing the course, students will have a strong background in programmable controllers, and they are prepared to go into industry and start writing programs immediately.

This book would also be quite helpful to a person in industry who needs to know more about programmable controllers. The main purpose of the text is to teach programming, which is accomplished by leading the reader through a systematic set of steps. It is designed to lead the user away from writing trial-and-error programs and toward using a structured method. Structured programs are much easier to follow, troubleshoot, and document. Finally, the text makes a good reference book. It is common to forget what is required if an engineer or technician does not use an instruction often enough. With this book, he or she can easily look up what is needed.

Acknowledgements

There are many people to thank. I would like to thank my editor, Charles Stewart, my assistant editor, Delia K. Uherec, and my production editor, Tricia Rawnsley. Without their help and patience, it would have been impossible for this textbook to come to fruition.

It is a pleasure to write with George. I would not have attempted to write this book without him. He fills in the gaps where I do not have sufficient knowledge, and he is a very good writer. I am lucky to have him as a collaborator.

I would like to thank my wife, Barbara, for proofreading the chapters I wrote. She has put many hours into fixing the grammar, formatting, and spelling. My daughter Sarah also proofread the manuscript for me. I felt quite comfortable asking her because she has many spelling trophies in her room.

My students were great at giving me feedback. I am lucky to have students who care. Michigan Technological University students are a great group to work with. My students call me after they get into industry and ask me to cover in class some material they found necessary in the work setting. I have used their recommendations extensively. There are three students in particular I would like to thank: Art Kilpela, Derek Christensen, and Chris Wells. All did considerable work on PID control and testing the information in the PID chapter.

Many thanks to the Prentice Hall reviewers, who offered many helpful suggestions on content and organization: John Golzy, DeVry Institute of Technology, Columbus, Oh.; W. Richard Polanin, Illinois Central College, Peoria, Ill.; William Mack, Harrisburg Area Community College, Harrisburg, Pa.; and Nebojsa Jaksic, DeVry Institute of Technology, Columbus, Oh.

Thanks also to Scott Hartz of Rockwell Automation/Allen Bradley for his support in getting equipment for my students' labs and technical photos for this book. I am sure I have overlooked many others who are not mentioned specifically— I wish to thank them too.

Robert Filer

Contents

PART 1

Programmable Controllers

1 Programmable Controllers

LEARNING OBJECTIVES

Upon reading this chapter students should be able to:

- Describe how PLC technology evolved from 1980 to 2001.
- Explain the difference between a PLC and a PC.
- Give the advantages of using a PLC.
- Explain the difference between sequential and automatic control.
- Define the basic terms used for PLC technology.

INTRODUCTION

This chapter looks at the history and evolution of programmable controllers. This evolution started with technicians using microprocessors to do simple repetitive sequential control. It quickly became obvious to users that microprocessors could perform these tasks accurately, with impressive economical advantage. When used in this fashion microprocessors need to have interfacing devices to go from digital to analog and vice versa. Eventually, this combination was packaged as a device called a programmable logic controller, or PLC. The number of tasks that a modern PLC can do today has multiplied rapidly as microprocessor technology has advanced. PLCs are good choices for many types of control problems, from simple discrete essential to complicated automatic control. Let' us take a look at this evolution.

▶ 1.1 HISTORY OF PROGRAMMABLE CONTROLLERS

Figure 1.1 shows how automatic sequential control has evolved from 1800 (when the relay was invented) to the turn of the 21st century. Relay control was dominant from 1800 until the late 1970s. The relay is not obsolete for power applications, where high voltage and current are needed, but for control applications its use has been greatly diminished. The demise of relay control started with the introduction of the integrated circuit (IC) in 1959. This invention became sophisticated enough, in about 10 years, to lower the cost of control while increasing complexity. The first programmable controller, designed by the Hydra-Matic Division of General Motors in 1968, used this technology. They had limited use, however, until Intel's introduction of inexpensive microprocessors in 1971.

The first PLCs were designed to replace relay control, which was used on assembly line production of cars. The changeover of tooling and control for automation, a very expensive process, was done every year a new model car was designed. The PLC significantly reduced this cost. Other industries soon found they could also benefit by changing their control, and today PLCs are used in everything from simple control in car washes to complex animation at major theme parks. There are PLCs in almost every processing plant. Today, the sales in programmable controllers has grown to be around $1 billion each year, and there are more than 30 manufacturers.

Personal-computer-based control is the PLC's latest competition, but PLC manufacturers are responding by designing software to emulate the PLC processor and by using rack mounted I/O to communicate with PCs. Personal-computer-based control has generated about a quarter of a billion dollars in annual sales, and PLC manufacturers have been successful in getting a large share of the control market by constantly adapting. Only the future will tell how long PLCs will remain competitive in control technology, but having a good track record for 30 years should help.

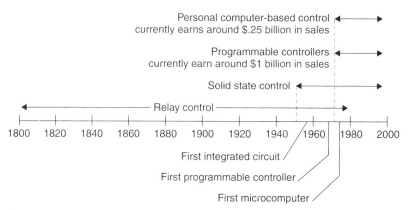

Figure 1.1 Evolution of programmable controllers.

▶ 1.2 PLC HARDWARE DEVELOPMENT

The evolution of PLC hardware has been fairly predictable because the PLC is, quite simply, a computer that has been manufactured to make it easy to interface with real-world control. Computer and *interfacing* technology are constantly changing, and manufacturers of PLCs must stay competitive with other control technologies. The first PLC made, in 1968, was large, slow, and limited by lack of memory and speed. Since then, PLCs have steadily become smaller, faster, and more powerful.

There are several sizes of PLCs available today. Figure 1.2 shows the range of PLC sizes relative to the number of I/O points available.

An interesting piece of information is that *nanos* are small enough to fit in your shirt pocket and have instruction sets that have about 90% of the capability of the largest PLCs. The cost of PLCs has also come down substantially, just as has that of personal computers. Both the nano and *micro* PLCs are small enough and inexpensive enough to make them great control devices for operations that are not too complicated. A nano or a micro could easily control a car wash or a farm's automated feeding station. They cost a few hundred dollars or less. At the other end of the spectrum are expensive PLCs with great flexibility and powerful commands. When there is enough demand for PLCs from a specific user, such as a plastics manufacturer, PLC manufacturers have made processors that are built for a specific application.

A PLC is a computer designed for control and interfacing so it has a completely different look from a desktop computer. A PLC has no keyboard, CD drive, monitor, or disk drive. Instead, it has a self-contained box with communication

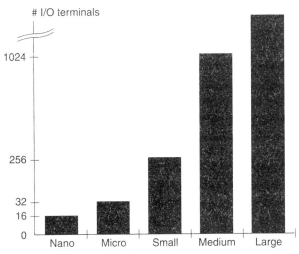

Figure 1.2 Range of PLC sizes relative to the number of I/O points available.

ports and a set of terminals for input and output devices. The box can be made to accept plug-in units for flexibility. The footprint for a PLC is usually much smaller than a typical PC. Figure 1.3 illustrates some Allen-Bradley PLCs.

Two major architecture styles of PLCs that have been developed are *fixed* and *modular.* Fixed architecture puts the processor and I/O in an enclosure that cannot be changed. This lowers the cost but limits flexibility. Nanos and micros come only in fixed architecture. This is because they are designed for relatively simple control applications that need a limited amount of I/O and because they must be able to compete in the marketplace and be priced competitively. Small PLCs come both in fixed and modular styles. A fixed unit has typically 12 to 24 input terminals and 8 to 16 output terminals with some add-on features. These units are cost-effective if you are controlling a process that does not change much and if you do not plan to move the PLC from one control process to another.

Small, medium, and *large PLCs* are all available with modular architecture, meaning users can buy components separately and put them in a *rack,* or *chassis.* This chassis, or rack, has a common bus to which modular units are connected when they are placed into it. The common bus provides power and gives the module and the processor access to each other's data. This architecture enables any user to customize PLCs for a wide range of simple to very complex applications. Once you buy a rack, you can add various power supplies, processors, and modules.

PLC processors have changed constantly, due both to demand from applications and to advancement in computer technology. Today, processors are faster and have more powerful instructions added as new models are introduced. This is very similar to the changes seen in PCs, starting with the first Apple computer in 1975, and continuing on today. Changes have been impressive, and PLCs use this same technology. Because PLCs are microcomputer-based, they can be made to do the things a computer can do. They are networked and can do tasks such as supervisory control and data acquisition *(SCADA),* as well as control. The newest PLCs have architecture that makes it possible to put more than one PLC processor in a rack, and networking is changing to accommodate smart devices connected to *Device Net.* Smart devices have their own microprocessor to let them run both dependent on and independent of the PLC.

The memory of the PLC is made to be nonvolatile so that if power is lost it holds its programming. When power then returns it can start controlling where it left off. The difference between an average PC and a PLC RAM is that the RAM on a PLC is made *nonvolatile* by a battery backup system or that its programming is written to *EEPROM* or EPROM. Once it is programmed, it is designed to be a stand-alone control unit and does not need any further attention in order to execute its programs.

Another useful feature of a PLC is that some are manufactured with a key so that the PLC can be put in the "run" mode and the key removed, preventing someone without the key from changing the PLC's programming.

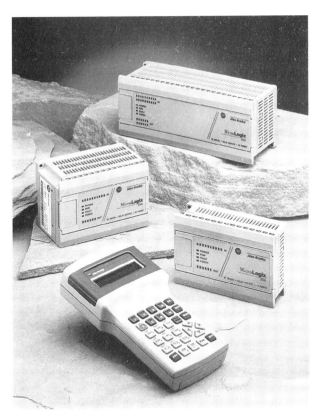

(a)

(b)

Figure 1.3 PLCs; (a) MicroLogix; (b) SLC500; (c) PLC-5; (d) ControlLogix.
Courtesy of Rockwell Automation/Allen-Bradley.

(c)

(d)

Figure 1.3 Continued

8

▶ 1.3 EVOLUTION OF PLC MODULES

The original PLCs were designed for sequential control only; however, due to demand, modules were added to do automatic control. At first modules were equipped with their own microprocessors to accomplish the automatic control task independently. This type of module is called a *smart module* because it can accomplish its task independent of the main PLC processors. This module architecture made it easy to plug another microprocessor-based controller into the rack for easy interfacing. Language modules enable programming in other high-level languages such as BASIC, but ladder-logic programming is by far still the most dominant form of programming.

Some of the types of modules that have been developed over the years are discussed next. Some of these modules, such as the PID, are no longer necessary because the newer PLC models include PID in their instruction sets.

Analog Input Module. The *analog input module* converts analog voltages or current signals to digital values and then stores this information for the processor to read. Many variations of this module have been developed to read different voltage and current levels.

Analog Output Module. The *analog output module* converts digital values read from the processor to analog voltages or current signals for devices connected to its terminal. Variations of this module have been developed with features such as high resolution, high speed, differential output, and single-end output.

ASCII Input Module. *ASCII input modules* convert *ASCII* code input information from an external device to alphanumeric information the PLC can process. The communication interfacing is done either through RS-232 or RS-422 protocol.

ASCII Output Module. *ASCII output modules* convert alphanumeric code from the PLC to ASCII code and output this to an external device such as a printer. The communication interfacing is done either through RS-232 or RS-422 protocol.

Barrel-Temperature Module. The *barrel-temperature module* enables users to monitor four zones of autotuned PID heating or cooling for temperature control. Molding machines and extruders commonly use this module for controlling barrel temperature while injecting material. The processor being used must be able to support this module.

BCD Input Module. *BCD input modules* read *BCD* from BCD devices such as thumb wheels and change it to uncoded binary.

BCD Output Module. *BCD output modules* write BCD to devices that require BCD codes.

Discrete Input Module. Discrete means only two states are allowed. *Discrete input modules* check the state a device is sending to its terminal and store it so the processor can later move the state to its input file. The state is stored in *binary* 0 if no voltage is detected and binary 1 if a threshold voltage is detected. Basically, the module changes the incoming voltage at its terminal to 5-V dc logic. The voltages at the terminals are isolated from the 5 V dc by using optoisolation. The many variations of this module can create different features, such as current-sinking dc input, current-source DC input, fast-response dc sinking input, current-sinking TTL input, current-source TTL input, and ac input.

Discrete Output Module. A *discrete output module* receives discrete output information from the processor and uses the information to send voltages to its terminals for attached operating devices. Basically, it receives 5-V dc logic information from the processor and converts the information to a specified voltage, which is directed to its terminals. Isolation is achieved through optoisolation. The numerous variations of this module can generate different features, such as current-sinking dc input ac output, high-current ac output, current-sinking dc output, current-source dc output, current-sinking TTL output, source TTL output, relay output, and electronically protected current-sourcing output.

Encoder Counter Module. Encoders keep track of the angular position of shafts or axes independently of the processor. This module lets the user read a relative or absolute encoder on a real-time basis and stores this information to be read later by the processor.

Gray Encoder Module. This module receives TTL *gray-code* signals from an input device and changes them to binary to be made available to the processor.

High-Speed Counter-Encoder Module. *High-speed counter-encoder modules* enable you to count and encode faster than you can with a regular control program written on a PLC, where the speed at which the control program can be executed is too slow. It has the electronics needed to count independently of the processor and to store the information or to send it out. The processor can access and manipulate its information through control programs. You need to check the frequency response of this module to make sure it is compatible with your needs. A common range available is from 0 to 50 KHz, which means the module would be able to count 50,000 pulses/s. Open-loop velocity-control modules are used to control hydraulic presses, weld-head placement, and die-casting machines and control the velocity of operation. An example is a press with a piston, which must be quickly accelerated to start the operation but slowed down as the press process is completed.

Isolated Input Module. *Isolated input modules* determine when dry contacts open and close, convert this information into digital, and store this information so the processor can read the digital information and place it in an input file.

Language Module. This module enables the user to program the processor using a language other than ladder logic, such as BASIC.

PID Module. *PID modules* enable the user to do proportional integral differential closed-loop automatic control. A set point and gains can be entered. Then, if properly tuned, the module will hold the process at the desired set point.

RTD. *RTD* stands for resistance temperature detector; it is used to measure temperature injunction with platinum, nickel, copper, and nickel-iron variable-resistance devices. Similar to the thermocouple module, it is designed to change the resistance read into scaled engineering units such as degrees Celsius or Fahrenheit and to work with PID instructions.

Synchronized Axes Module. A *synchronized axes module* provides logic to synchronize multiple axes. Examples of applications are hydraulic tailgate loaders, forging machines, and roll-positioning operations. A mold-pressure module is connected to strain gauges to detect mold-cavity and hydraulic pressure. This information is used in a control program to improve repeatability.

Thermocouple Module. *Thermocouple modules* interface with type J, K, E, R, S, B, C, and D thermocouples, but not all models have all these types. You need to be sure the thermocouple you are using is covered by the model you buy. These modules enable you to read millivolt signals that standard analog modules cannot read. They are designed to change the millivolt signal into scaled engineering units such as degrees Celsius or Fahrenheit and to work with PID instructions.

Thumb-Wheel Module. A *thumb-wheel module* reads a TTL BCD thumb-wheel device in parallel and changes it to uncoded binary for the processor to read.

TTL Input Module. *TTL input modules* read voltages produced by TTL devices and make this information available to the processor.

Modules will continue to evolve to meet specific application demands. The modular architecture will also continue to be competitive because it is so flexible and easy to use.

▶ 1.4 EVOLUTION OF PLC SOFTWARE

Software for programming PLCs has evolved in an interesting fashion. The first PLCs used a special language written to make it easy for the people who were creating relay logic control to program the PLC. Ladder-logic programming language uses graphic symbols that show their intended outcome instead of words. This automatically gives a visual, or graphic, presentation of the control. Displaying information graphically often changes something incomprehensible into information that is almost instantly comprehensible. A good example of this is using the reams of temperature information from weather stations all over the world into a weather map showing the temperature by colored area for each range of temperature. Such a weather map shows information much more clearly.

Ladder logic is still the most common means of programming PLCs, even though other languages are made available by PLC manufacturers. It has remained the dominant programming language for more than 30 years because of the graphical presentation. When you are troubleshooting control, the software can easily show what is happening in real time by highlighting the instructions for true or active inputs and outputs. Once you are familiar with ladder logic, you can watch what is going on in real time on the monitor screen and readily diagnose problems. Using a structured approach when writing the control programs greatly enhances the graphical display of the process being controlled. Ladder logic will remain dominant until a better graphical language evolves. Included in this book is Appendix A, *Relay Logic,* which shows how relays were used in control prior to PLCs. A review of relay logic is helpful in understanding the ladder-logic programming.

A revolution in graphical programming is occurring, in which a subroutine is shown as a functional control block rather than showing the ladder logic. A ladder-logic program must be written for the control block, but it is not visible. The overall control is shown graphically by connecting the functional control blocks together. Rockwell International RS Logix Frameworks software is an example of this. PLC programming will continue to evolve and change, which creates a dilemma. The more flexibility you build into a computer-control device such as a PLC, the more you have to know and the harder it becomes to use the device. Simple PLCs are easier to use but less powerful, whereas more advanced ones require the user to know much more.

There is a movement to standardize the programming of PLCs so that all the manufacturers have the same programming language. However, dominant manufacturers can lose market share by doing this. Consequently, standardizing has begun but has not reached the point where it is generally used. You must be sure the generic software available is compatible with your particular PLC.

▶ 1.5 KEY TERMS AND DEFINITIONS

Every field of technology has its own special terms, and part of learning that technology is becoming acquainted with its terminology. The following are some terms to become familiar with.

Automatic Control.   A process where feedback is used to continuously check for errors and to adjust an output to a desired point.

Discrete Control.   Control that allows only two states for inputs and outputs. The two states are usually on and off.

Ladder-Logic Programming.   A special language written to make it easy for the people who were creating relay logic control to program the PLC. *Ladder-logic programming* language uses symbols rather than words to show what they are intended to do. Most PLCs are programmed with this language.

PLC.   An acronym for programmable logic controller. A PLC is a microprocessor-based device with built-in interfacing that can be programmed to automatically control a process. It is really a computer package for ease of interfacing that is designed to work in industrial environments. PLCs are often referred to as programmable controllers rather than programmable logic controllers, but the acronym PC is widely used to refer to personal computers, so retaining the L differentiates between the two thus avoids confusion.

PLC Module.   A device that plugs into a PLC rack or chassis and enables it to get power and communicate with the processor.

PLC Processor.   A computer designed specifically for programmable controllers. It supervises the action of the modules attached to it.

SCADA.   An acronym that stands for supervisory control and data acquisition. This module is used to control and interrogate a device remotely.

Sequential Control.   A process that dictates the order of the steps and the completion of one step before the next step is initiated.

1.6.1 Ease of Use and Programming

PLCs were specifically designed for control, and easy programming and interfacing were major considerations. It is hard to beat something that is specifically designed for what you want to do. The PLC is literally a custom-made tool for accomplishing control of a task. Its successful design and use have generated yearly sales of billions of dollars, which is significant for a niche market. PLCs were initially designed for sequential control, but—because of their success—users have asked for modules and instructions for automatic control.

1.6.2 Cost Effectiveness

The original PLCs were designed to replace relay control, and the cost savings were so obvious that relay control is becoming obsolete except for power applications. Today PLCs are mass-produced in large quantities by more than two dozen U.S. manufacturers, and they are also manufactured internationally. Their prices have fallen just as have PC prices. A modern PLC is cheaper and more powerful than its predecessor. One area where PLCs are most cost effective is in small control applications. For a few hundred dollars, you get a computer controlling your process. Assembly-line shutdowns can cost several thousand dollars an hour, and the ease of troubleshooting a PLC program can significantly reduce downtimes.

1.6.3 Reliability

There are millions of PLCs in service and their track record has been excellent. Many of them have been in service for more than 20 years. If they were not reliable, they would not be used on such a wide scale.

1.6.4 Flexibility

Because they are really computers, a PLC's flexibility is one of its strongest points. A PLC can be moved from one application to another simply by reprogramming it and changing some modules. The cost involved is usually minimal.

1.6.5 Ease of Troubleshooting

There are two reasons PLCs are easy to troubleshoot: (1) The control program can be displayed on a monitor and watched in real time as it executes, and (2) a PLC can be reprogrammed immediately to fix problems. The modular architecture also makes it possible to replace a bad component quickly by removing the component and replacing it with a new one. Structured programming is available to the user; if used, it makes the ease of troubleshooting significantly better.

1.6.6 Availability of Training and Technical Support

The major manufacturers all offer training courses throughout the major cities in the country, or they will do on-site training. This is important if you are using computer-based technology. Good technical support is also available via phone or e-mail.

SUMMARY

We have seen that programmable controllers have constantly changed and advanced with the microprocessor technology. The tasks that they can accomplish reliably and economically are impressive and make them serious contenders as a means of implementing discrete sequential as well as automatic control. The PLC business generates billions of dollars in sales, so there is great incentive to keep it evolving so it can continue to compete. It is well worth your time to learn about this savvy way to solve control problems.

▶ ## EXERCISES

1. Discuss the physical differences between a PLC and your desktop PC.
2. How does the PLC's memory differ from your PC's memory?
3. What is the most popular language for programming PLCs and why is it usually chosen over others?
4. What is the difference between sequential control and automatic control?
5. Give some reasons for networking PLCs.
6. What makes a module smart?
7. What makes a PLC cost-effective?

2 | Interfacing and Ladder-Logic Fundamentals

CHAPTER OUTLINE

LEARNING OBJECTIVES

Upon reading this chapter students should be able to:

- Explain how ladder logic gets its name and how it became the language of choice for PLC programming.
- Write the ladder-logic programming to create the following logic gates: inverter, AND, OR, and Exclusive OR.
- Draw the electronics for a simple I/O module and explain how it works.
- Show how to change easily from binary to hex and octal and vice versa without a calculator.
- Convert positive or negative numbers into signed binary numbers and vice versa.
- Explain how PLCs scan.

INTRODUCTION

This chapter is intended to give you the basic information needed to understand how a PLC interfaces with external devices. Then, the chapter shows how digital gates are duplicated in ladder logic using simple PLC programs. Last, it discusses number systems used in PLCs and scanning. This will enable you to become familiar enough with ladder-logic programming to get started writing and analyzing simple programmable controller programs.

▶ 2.1 INTERFACING WITH THE PLC

PLC control is accomplished as it receives information from various devices so that it can initiate changes and also as it sends out information to devices that can cause changes. When you use a PLC for control, you need a way to receive and send information to external devices. The external devices are usually not computers and are operating at voltages other than 5 V dc; therefore, an intervening device must transfer information to the computer as well as send out electrical signals to the external devices.

This is where plug-in modules come into play. The processor and modules connect to a motherboard in the back of the rack that contains the data, address, and control buses. Communication between the processor and the modules then takes place via these buses. The module can have incoming or outgoing signals. One purpose of the module is to change external incoming signals on its terminals to digital signals that are compatible with the data bus of the processor. Another purpose is to change digital data from the processor to usable signals that drive external devices connected to its terminals. Input modules must have the electronics for changing the incoming voltages from external devices to voltages a computer port can read. Figure 2.1 shows an example of how this might be done for an input module.

The incoming voltage from a push button goes to a screw terminal in the front of the module. The electronics inside rectify, filter, and change the voltage level to

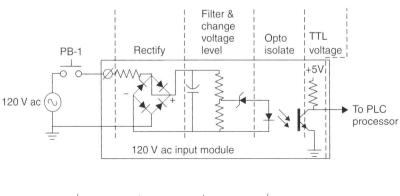

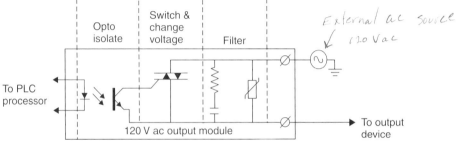

Figure 2.1 Example circuit for I/O modules.

one compatible with the bus to which it is attached in the rack. In addition to a change in voltage, electrical isolation between the external device and the PLC is necessary. The PLC is a computer that is vulnerable to electrical signals from the outside world, particularly voltages that power motors and similar devices. The isolation is achieved via *optical isolation* by using an LED to turn a transistor on and off, thus sending digital signals to the PLC. A TV remote control illustrates optical isolation. If you are operating a remote control, you cannot receive an electrical shock, no matter what is happening with the TV.

Figure 2.1 also shows an example of how this might be done for an output model. The PLC sends a signal to the output module via the back connector, and the electronics inside isolate, filter, and change the voltage level to one compatible with the device connected to a terminal in the front of the module.

The module must also have electronics circuitry for communicating with the PLC. The PLC will issue read/write control signals to the module, and the module needs to respond as directed.

Figure 2.2 shows some examples of external devices connected to input and output modules. This is a very limited group of devices, and many other types of

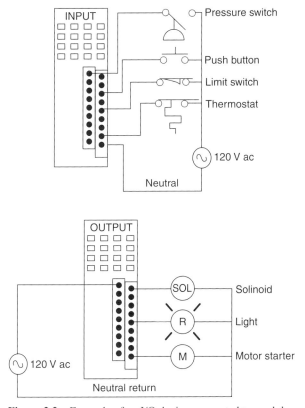

Figure 2.2 Example of an I/O device connected to modules.

devices are possible. The list of modules given in Chapter 1 will give you a better idea of what is possible.

▶ 2.2 PLC SCANNING

When a PLC executes a program, it must know—in real time—when external devices controlling a process are changing. It accomplishes this by constantly scanning the inputs. Figure 2.3 shows a graphical snapshot of this process. First, the PLC reads the inputs, and then it executes the program depending on these inputs. Next, it updates the outputs as the program dictates; finally, it constantly repeats this cycle as long as the PLC is in the run mode. When executing a program, it starts at the top and works its way through the rungs, one rung at a time, in the order they are written. It will not deviate from this order unless given special instructions designed to change the order.

▶ 2.3 LADDER LOGIC

Figure 2.4 shows a control program for a programmable controller. Even though you have not done any programming, you can learn some basic *instructions* and some ladder-logic concepts. First, you can see from the general appearance of the program why it is called ladder logic. There are two outside, parallel vertical lines on the left and the right; between these are horizontal lines with symbolic programming. The general outline shape looks like a ladder with rungs. Each line of horizontal programming is called a rung. The first rung on this ladder is rung 2:0, where 2 designates the file number in which the program is located and 0 is the rung number.

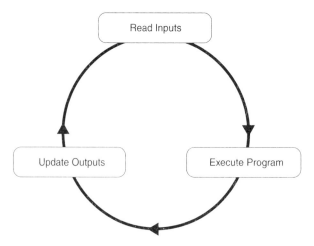

Figure 2.3 PLC scanning.

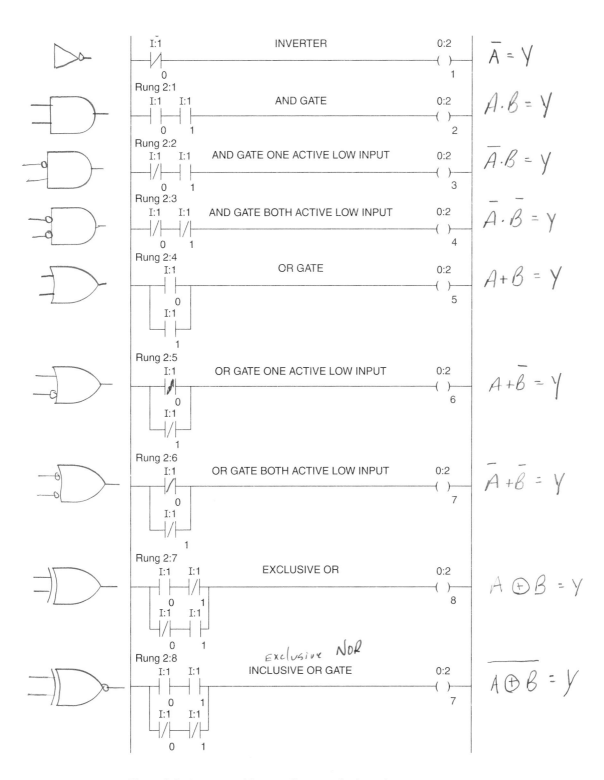

Figure 2.4 Programmable controller control schematic.

21

The next question is, where does the logic come from? You need to understand three symbols in this program to see the logic. The first symbol is $-]$ $[-$, an instruction that tells the processor to look at a specific bit in its RAM memory. If the bit is 1, the instruction is true, and if it is 0, the instruction is false. The determining factor in choosing which bits in its memory to look at is the address. The address I:1/0 indicates file I, element 1, bit 0. I stands for input file, and the PLC automatically sets up this file to keep track of inputs to the terminals of the input modules. The 1 tells which slot in the rack the module is in, and the 0 tells the terminal. The PLC constantly checks the terminals of the input module and puts that information in a special input file. So, in this case, the instruction tells the PLC to check the bit set up for keeping track of terminal 0 of the input module in slot 1. The module will tell the processor the bit should be 1 if the correct voltage is present at the terminal; if it is not present, the module will tell the processor to set the bit to 0.

The second instruction, $-]/[-$, plays the same role as $-]$ $[-$, except that if the bit addressed is 1, the instruction is false and if it is 0, the instruction is true. The third symbol, ( ), is for outputting to the output module. If the instructions to the left on its rung have a true path to the leftmost vertical rail, then the PLC will set the bit to which it is referenced via the address to 1. If no complete true path is available, it will set the bit to 0. The processor constantly updates the terminal of the output module to tell it when it should send out a voltage to one of its terminals. The address O:2/5 tells the output module in slot 2, terminal 5, to turn on if the bit is 1 and off if 0.

Let us assume that we have an input module in slot 1 and have two push buttons connected to terminals 0 and 1, respectively. Each push button has two terminals, one connected to 120 V ac and the other connected to either terminal 0 or 1 of the input module. Pushing the button results in 120 V ac being applied to the module terminal to which it is connected. Also, there is an output module in slot 2. Given this example, let us look at the rungs of instructions in Figure 2.4 and see what would happen.

Rung 2:0   If we *do not* push the push button connected to terminal 0, no voltage will be present, and when the module is read, the processor puts a 0 for the bit in the input file set up for it. Then, when the processor executes this rung via the input instruction]/[, it sees the rung's input condition as true. Because the input condition is true, the processor puts a 1 in the output file for the bit that is addressed, as directed by the output instruction ( ). Finally, the processor tells the output module terminal 0 to turn on, and to do this the processor updates the output module. This causes the output module in slot 2 to turn on terminal 0, and terminal 0 outputs 120 V ac. The type of input instruction,]/[, in this rung inverts because it causes the output to be in the opposite state of the input. No action (push button not pushed) at the input results in action because the output terminal goes on.

Rung 2:1. The processor works through the inputs from left to right, and the output instruction sees a true condition only when it sees a true path via input instructions to the left rail. If you push both push buttons connected to terminals 0 and 1, respectively, 120 V ac will be present at the terminal, and the input module will tell the processor to put a 1 in the bit setup for each input instruction. Using this instruction, the processor says both instructions are true and sets the bit to 1 for the output module in slot 2, terminal 1. The output module is updated, sees the 1, and causes terminal 0 to go to 120 V ac. If none or one push button is pushed, the input conditions to the left of the output instruction do not produce a true path. This type of program rung is called an *AND* because both push buttons, 0 AND 1, must be pushed to activate the output.

Rung 2:2. If you *do not* push push button 0 but *do* push push button 1, the processor uses the input instructions to say both instructions are true and sets the bit to 1 for the output module in slot 2, terminal 1. The output module is updated, sees the 1, and causes terminal 0 to go to 120 V ac. All other combinations result in no true path to the left rail. This type of program rung is an AND with one input active low, because no action is required for push button 0.

Rung 2:3. If we *do not* push both push buttons connected to terminals 0 and 1, respectively, 120 V ac is not present at the terminals, and the module tells the processor to put a 0 in the bit setup for each. The processor uses this instruction to say that both instructions are true and sets the bit to 1 for the output module in slot 2, terminal 1. The output module is updated, sees the 1, and causes terminal 0 to go to 120 V ac. All other combinations result in no true path to the left rail. This type of programming is an example of an AND function, because both push buttons 0 AND 1 must not be pushed. This is an AND gate with active low inputs, because no action for the input is required to turn the output on.

Rung 2:4. This type of instruction is an *OR* gate because there are two paths that will turn the output: either push button 0 OR 1 is pushed.

Rung 2:5. This type of instruction is an OR gate with one input active low because there are two paths that turn on the output: either push button 0 is not pushed OR 1 is pushed.

Rung 2:6. This type of instruction is an OR gate with both inputs active low because there are two paths that turn on the output: either push button 0 OR 1 is not pushed.

Rung 2:7. This is an *Exclusive OR gate* because there are two paths that turn on the output: either push button 0 is not pushed and 1 is pushed OR push button 0 is pushed and 1 is not pushed. The push buttons have to be in one of two opposite states to turn on the output but cannot be in the same state. If they are simultaneously pushed or not pushed, the output will not turn on.

Rung 2:8.   This is an *Inclusive* OR *gate* because there are two paths that turn on the output: either both push buttons are not pushed OR both push buttons are pushed. They both must be in one of the two states to turn the output on. If both buttons are simultaneously pushed or not pushed, the output will turn on.

The availability of these rungs for creating gates via ladder-logic programming creates powerful decision-making capability. A microprocessor chip is loaded with millions of these gates, but they are made as hardware rather than software by using transistors instead of programming. It takes millions of these transistors to make the necessary gates inside a modern microprocessor, but this is no problem for chip makers. This capability could be duplicated in ladder logic, but it would take millions of rungs of ladder-logic programming, which is not practical. Once you know these gates, you have the tools to do complicated programming. Figure 2.5 shows the symbols used for these gates in digital form and the equivalent ladder logic. Obviously, control done in ~~digital~~ *ladder logic* can be duplicated in digital and vice versa.

We discuss one last useful introductory program. Figure 2.6 illustrates a means of detecting if an output has been operated, called a *seal-in circuit.* It works like this. If on a *scan* I:1/0 goes true, then output O:2/5 goes true and its *bit* in memory is set to 1 when the output is updated. On the scan following that input, O:2/5 sees a 1. Even though input I:1/0 goes false, output O:2/0 stays on because it has a true path to the left rail via its own bit. The only way to turn off output O:2/0 is to push a reset. Note in the scan cycle that inputs are always read first, which is why the output cannot be turned off. Once the input O:2/5 is set to 1, there will always be a true path to keep the output on.

▶ 2.4 TRANSFERRING A PROGRAM TO PLC

The instructions for your ladder-logic control program are generated by using a software package for the processor you are using. Ultimately, the software will have to communicate with the processor in a language that it understands, and that language is called *machine language,* or *binary.* The processor, being a digital device, understands only 0s and 1s. Figure 2.7 shows how ladder logic is changed to machine language. The program you develop is sent to an interpreter, or assembler, which converts it to machine language for the processor. The processor reads this machine language by putting it in a file. A special file, made up of words that are 16 bits or 32 bits long, is set aside by the processor. The processor can then execute this program when told to do so by being put in the run mode.

	Ladder Logic	Digital Symbol	ANSI/IEEE Symbol

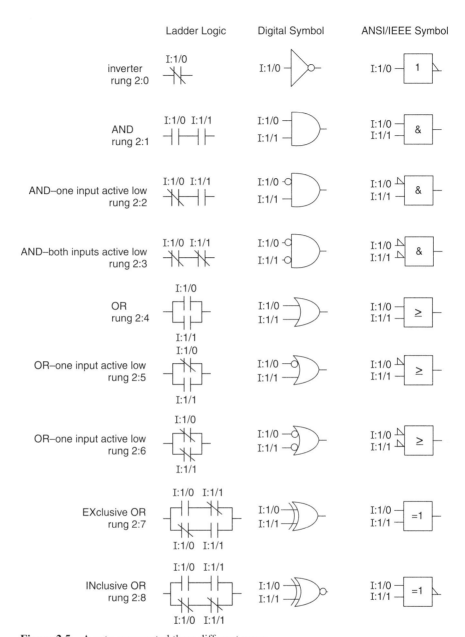

Figure 2.5 A gate represented three different ways.

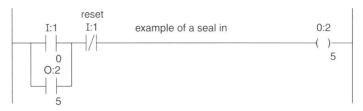

Figure 2.6 Seal-in circuit.

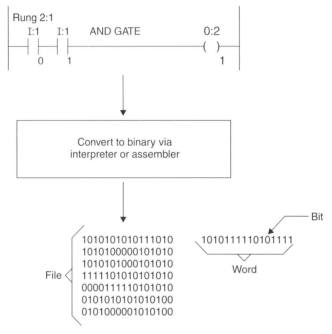

Figure 2.7 Ladder logic changed to machine language.

► 2.5 NUMBER SYSTEMS

Sometimes when you are programming, you will be required by the instruction to put in binary information for filtering so that each bit in a word is treated differently, depending on whether the filter has a 0 or a 1 in its position. It is easier to enter binary information in hexadecimal. Hexidecimal helps eliminate the problem of making a mistake when entering a long string of 0s and 1s. It is easy to convert binary numbers to hexadecimal and hexadecimal numbers to binary. Table 2.1 shows how to do this.

TABLE 2.1 NUMBER CONVERSION

Decimal	3-bit binary	Octal	4-bit binary	Hexidecimal
0	000	0	0000	0
1	001	1	0001	1
2	010	2	0010	2
3	011	3	0011	3
4	100	4	0100	4
5	101	5	0101	5
6	110	6	0110	6
7	111	7	0111	7
8	000	10	1000	8
9	001	11	1001	9
10	010	12	1010	A
11	011	13	1011	B
12	100	14	1100	C
13	101	15	1101	D
14	110	16	1110	E
15	111	17	1111	F

Binary to octal	**Octal to binary**
1011011110101111	1 3 3 6 5 7
↓	↓
001 011 011 110 101 111	001 011 011 110 101 111
↓	↓
1 3 3 6 5 7	1011011110101111

Binary to hex	**Hex to binary**
1011011110101111	B 7 A F
↓	↓
1011 0111 1010 1111	1011 0111 1010 1111
↓	↓
B 7 A F	1011011110101111

Decimal, *octal*, and *hexadecimal numbers* can be converted to 3- and 4-bit code. To convert a binary number to octal, you break the binary number into groups of three, starting on the right and working left. If the binary number is not divisible by three, simply add leading 0s to complete the final group. This does not change the binary number's value. Then you substitute the equivalent octal value for each 3-bit group of binary digits. The resulting octal number is equivalent in value to the original binary number. Reversing the process involves changing the octal digits to equivalent 3-bit codes and making the groups into an unbroken string. You use exactly the same process for going from hex to binary or binary to hex, except the binary codes are 4 bits instead of 3.

▶ 2.6 SIGNED INTEGER NUMBERS

One way for the PLC to distinguish positive and negative numbers is to use the *most significant bit* (MSB). A 0 indicates a + and 1 indicates a –. Figure 2.8(a) shows 16 bits can represent numbers ranging from +32,767 to –32,768. Negative numbers are not stored in true binary form; instead, they are stored in 2's complement because it is easier for the computer to perform mathematical operations when this form is used. The correct sign bit can be generated by forming the 2's complement given the true binary value. When you enter a –3 using the keyboard, the – key tells the computer to store the number in 2's complement. The computer does this by first taking the 1's complement of the binary number of the decimal key pushed, which is done by simply changing 1s to 0s or the 0s to 1s on a bit-by-bit basis. Then the 2's complement is formed by adding 1 to the 1's complement. Forming the 2's complement automatically places a 1 in the MSB; when the PLC later retrieves that number from memory, the 1 in the MSB tells the PLC that the number is negative.

Figures 2.8(b) and (c) shows how –1 and –32,768 are changed to 2's complement. Interestingly enough, changing a number in 2's complement form

```
     0111111111111111    +32767
          ↓         ↓
     0000000000000011    +3
     0000000000000010    +2
(a)  0000000000000001    +1
     0000000000000000    +0
     1111111111111111    −1
     1111111111111110    −2
     1111111111111101    −3
          ↓         ↓
     1000000000000001    −32767
     1000000000000000    −32768
```

(b) 0000000000000001 → 1111111111111110   Complemented
                                    +1
                   1111111111111111   2's complement representing −1

(c) 1000000000000000 → 0111111111111111   Complemented
                                    +1
                   1000000000000000   2's complement representing −32768

Figure 2.8 Signed integer numbers; (a) (b) unsigned binary for –1; (c) unsigned binary for 32768.

back to unsigned binary form is done the same way as changing an unassigned binary number to 2's complement form. You simply take the 1's complement and add 1.

SUMMARY

Now that you understand how a PLC works and how the basic concepts given in this chapter form the foundation for writing more complex programs, you are ready to learn more detail to enable you to become a proficient programmer. Modern PLCs have powerful instruction sets, and the next chapters will take you through the details of using these instructions and features.

▶ ## EXERCISES

1. Show the programming required for the following task:

   A 2-ton press is used to make fenders for a car. A sheet of steel is formed into the proper shape. An operator feeds the sheet into the press and then, for safety reasons, must press two buttons simultaneously, one with the left hand and the other with the right hand. The buttons are located so the operator's hands are out of the path of the press. A light comes on and a foot pedal then causes the press to operate. The foot pedal must not work until the light for the hand indicator comes on.

2. Show the programming required for the following task:

   A jet fighter has an alarm panel which will initiate an alarm from three contacts, which are closed by detectors. The detectors are: out of fuel, missile coming, and obstacle ahead. Any one of these or any combination should cause an alarm.

3. Two push buttons have to be pushed in the right order to operate a door. The order is 1 and then 2. If the right sequence isn't pushed, an alarm is to go off and keep the door from operating until a reset button is pushed. Show the programming required.

4. What does the term *interfacing* mean?

5. Without using a calculator, change the following binary numbers to hex.
   a. 1110111110110101
   b. 1010110110110001
   c. 1011110110110101
   d. 1111110110111001

6. Without using a calculator, change the following hex numbers to binary.
   a. AF5C
   b. EDF6
   c. 987A
   d. 5CB6

7. Assume the most significant bit is the sign bit. What negative number is represented by 1111000011110000?

8. Convert –33 to 16-bit 2's complement.

9. Does the order in which you write the rungs of a ladder-logic program matter? Explain.

3 SLC 500 Processor Architecture versus ControlLogix

CHAPTER OUTLINE

3.1   SLC 500 Files
3.2   ControlLogix Memory Layout
3.3   Project Organization
3.4   ControlLogix Data File Types
      Summary

LEARNING OBJECTIVES

Upon reading this chapter the students should be able to:

- Identify SLC 500 program files.
- Identify SLC 500 data file types.
- Organize and apply the various SLC 500 data table types.
- Define tasks, programs, and routines in a ControlLogix controller.
- Identify ControlLogix data file types.
- Recognize the difference between user-defined and predefined data file types.
- Organize and apply the various ControlLogix data file types.

INTRODUCTION

This chapter identifies the different file types in SLC 500 and ControlLogix processors. The organization and application of program files and data files are covered for both types of processors. The different data file types are explained, along with how to determine which file type to use in an application and how to organize the files.

► 3.1 SLC 500 FILES

There are two basic types of files stored in the SLC 500 controller, *program files* and *data files*. The program files store ladder logic and the data files store the data accessed by the ladder logic.

When a project is created for the SLC 500, there are default files created for both the program files and the data files. Additional files may be created as needed.

3.1.1 Program Files

Program files 0, 1, and 2 are the default program files created when a project is created.

Program file 0 is the system file. It is one of the two files that does not contain ladder logic. It stores passwords, the processor type, I/O configuration, and the processor project name.

Program file 1 also does not contain ladder logic. It is a reserved file.

Program file 2 is the first file to contain ladder logic and is the main control program file. It is a required file and is necessary to run any additional program files.

Program files 3–255 are subroutine files, which must be created. They may be regular subroutine files accessed by Jump to Subroutine instructions, or they may be designated subroutine types, such as a Selectable Timed Interrupt, High-Speed Counter Interrupt, or User Error Fault routine.

Using multiple program files is a common programming technique. There are a number of reasons to use subroutines.

A subroutine may contain ladder logic to control a portion of a process that is not always needed. That subroutine may be executed only when necessary, which saves on scan time when it is not being executed.

Different sequences of operation may be necessary on a machine when different parts are made on the same machine. The sequences for the various parts may then be stored in separate subroutines. Then it is necessary to execute only the subroutine necessary to build a specified part.

It is also very common to use subroutines to break a project into smaller pieces. Instead of putting all the ladder logic in a few files, you may break up the program into smaller functional groups. You may then place the logic of these functional groups in separate subroutines. This breaks up a large program into smaller components, results in better program organization, and simplifies the troubleshooting process.

3.1.2 Data Files

Data files, which make up a data table, store the data accessed by the ladder logic. Data are accessed at the bit level, word level, or element level. A bit is a binary digit. It can be either 1 or 0. A *word* consists of 16 bits. An *element* may consist of a single word or multiple words, depending on the data table type.

When a project is created, there are nine default data files created. The default data files are as follows:

File 0: The *output file* stores the data that are sent to the output modules. The output file is created by configuring the output modules using I/O configuration.

File 1: The *input file* stores the data received from the input modules. The input file is created by configuring the input modules by using I/O configuration.

File 2: The *status file* contains information regarding processor status. Communication channel information, processor fault information, and scan times are some of the examples of data stored in the status file.

File 3: The *binary file* is used for internal relay storage and for storing data when you want to have the default display in bit status. It is also referred to as the *bit file*.

File 4: The *timer file* stores the timer's status bits, preset values, and accumulated values.

File 5: The *counter file* stores the counter's status bits, preset values, and accumulated values.

File 6: The *control file* stores the control's status bits, length values, and position values. It is used as the control in many file-level instructions.

File 7: The *integer file* is used to store whole numbers and bit status. The default display is in a decimal format.

File 8: The *floating point file* is used to store fractional numbers and numbers with a range of 1.1754944e-38 to 3.40282347e+38. It cannot be used for individual bit storage.

Files 9–255: *User defined files*

Output Data File. The output data file stores the status and values sent to output modules. The file is created when the output modules are configured in I/O configuration. The format for an output address depends on the level. The word-level format is O:slot number of module location.word number. For example, O:2.1 represents word 1 of an output module located in slot 2.

The bit-level format is O:slot number.word number/bit number (0–15). For example, O:3.0/5 represents bit 5 in word 0 of an output module located in slot 3.

If 32-point modules were used, they would require 32 bits, or two 16-bit words. The address for the 32-point module would be words 0 and 1. For example, for an output module located in slot 4, the addresses would be O:4.0/0–15 and O:4.1/0–15.

Input Data File.   The input data file stores the status and values read from input modules. The file is created when the input modules are configured using I/O configuration. The format for an input address is as follows:

The word-level format is I:slot number of module location.word number. For example, I:1.1 represents word 1 of an input module located in slot 1.

The bit-level format is I:slot number.word number/bit number (0–15). For example, I:4.0/5 represents bit 5 in word 0 of an input module located in slot 4.

If 32-point modules were used, they would require 32 bits, or two 16-bit words. The address for the 32-point module would be words 0 and 1. For example, for an input module located in slot 5, the address would be O:5.0/0–15 and O:5.1/0–15.

Status Data File.   The status file, file 2, contains data on processor status. The status file is created when the project is created, and its size cannot be modified by the programmer.

The status file contains information such as a real-time clock, processor type, scan-time data, math status bits, I/O slot status, channel communication data, debug data, fault information, and other processor status data.

Addresses for the status file start with the letter S, followed by the element number and bit number if required. For example, the address for a free-running clock is S:4.

Binary Data File.   This file is also referred to as the bit file. File 3 is the default file for the binary data file. Binary files can also be assigned file numbers 9–255. Each file number must be a unique type.

The binary file is used to store internal relay status and to store data when it is desirable to have a default binary display. The address format is

B3:element/bit number (if required)

For example, the element-level format is B3:5 (binary file 3, element 5). At the bit level, there are two formats for displaying addresses:

B file number:element number/bit   or   B file number/bit

The addresses B3:1/1 and B3/17 are equivalent addresses.

Timer File. The timer file stores the timer's status bits, preset values, and accumulated values. A timer is a three-word element. The default file for timers is file 4. Timers can also be assigned to files 9–255. The range of values that can be stored in the preset or accumulated word is 0–32,767.

The following is an example of a timer's addresses:

Status bits: T4:0/EN (enable bit), T4:0/TT (timer timing bit), T4:0/DN (done bit)

Preset value: T4:0.PRE

Accumulated value: T4:0.ACC

Counter File. The counter file stores the counter's status bits, preset values, and accumulated values. A counter is a three-word element. The default file for counters is 5. Counters may also be assigned to files 9–255. The range of values that can be stored in the preset or accumulated word is 0–32,767.

The following is an example of a counter's address:

Status bits: C5:0/CU (count-up enable bit), C5:0/CD (count-down enable bit), C5:0/DN (done bit)

Preset value: C5:0.PRE

Accumulated value: C5:0.ACC

Control File. The control file stores the data associated with the control element. The default file for controls is 6. Files 9–255 can also be assigned to the control type. This element uses three words to store the status bits, the length, and the position (a three-word element). The control element is used for control of file-level instructions, such as sequencers and shift registers.

The following is an example of a control element's addressing:

Status bits: R6:0/EN (enable bit), R6:0/DN (done bit), R6:0/ER (error bit)

Length word: R6:0.LEN

Position word: R6:0.POS

Integer File. Integer files store values used in the program. The default file for integers is file 7. Files 9–255 can also be assigned to the integer type. It is a single-word element, so it uses 16 bits to store values. Positive numbers are stored in the first 15 bits in a binary format, with bit 16 being the sign bit. Negative numbers are stored in 2's complement binary. The numbers $-32,768$ through $+32,767$ can be stored at an integer address.

An example of an integer address is the following:

Integer word: N7:20 (file 7, element 20)
Integer bit: N7:20/10 (bit range is 0–15)

Floating Point. Floating-point files also store values used in the program. They store fractional, very large, and very small numbers. The default file for floating point is file 8. Files 9–255 can also be assigned as floating-point files. In the SLC 500 processors, the following can use floating-point data files:

5/03 OS301, OS302,
SLC 5/04 OS400, OS401
SLC 5/05

OS represents operating system. In the fixed, 5/01 and 5/02 processors, data file 8 is reserved.

The floating-point data element is two words or 32 bits. The value range that can be stored in a floating-point element is $1.1754944e^{-38}$ to $3.40282347e^{+38}$, 0, and $-3.402823e^{+38}$ to $-1.1754944e^{-38}$.

An example address for a floating-point element is the following:

F8:20 (file 8, element 20)

User-Defined Files 9–255. User-defined files are files 9–255. The file types that can be created in all SLC 500 processors are bit, timer, counter, control, and integer. In SLC 5/03, OS301, OS302; 5/04, OS400, OS401; and SLC 5/05 processors, Floating Point, String (ST), and ASCII (A) data types can be created. Each file number is a unique file type.

▶ 3.2 CONTROLLOGIX MEMORY LAYOUT

A *project file* is created for the ControlLogix controller that contains the programming and the configuration information. The project file contains the following defined properties:

- Chassis size/type
- Slot number of the controller
- Description (optional)
- File path
- Project name (file name)
  - The file name is also assigned to the controller name.
  - The controller name can be changed.
  - When the file is saved under a different name, the controller name does not change.
  - The file name consists of 40 characters, made up of letters, numbers, or underscores.
  - The file name must begin with a letter or an underscore.
  - The file name cannot have consecutive or trailing underscores.

► 3.3 PROJECT ORGANIZATION

A project consists of three major components, *tasks, programs,* and *routines.* Tasks configure the controller execution, programs group logic and data, and routines contain the logic of the executable code.

The controller operating system is a preemptive multitasking system in that it has the ability to interrupt an executing task, switch control to a different task, and then return control to the original task once the interrupting task completes its execution. Only one task may be executing at any given time and only one program within a task may be executing at any given time.

3.3.1 Tasks

Tasks schedule and provide priority for executing programs or a group of programs assigned to the task. There are two types of tasks, continuous and periodic. The *Logix5550* controller supports up to 1 continuous and 32 periodic tasks. If there is no continuous task, there can be 32 periodic tasks, but if there is a continuous task, then there can only be 31 periodic tasks.

Each task has an assigned priority level to determine which task executes if multiple tasks are called. There are 15 priority levels that can be assigned to periodic tasks, with 1 being the highest priority and 15 being the lowest priority. A higher-priority task interrupts a lower-priority task. A continuous task has the lowest priority and is always interrupted by a periodic task. The continuous task restarts itself after each execution, but it may be interrupted by a periodic task. When a project is created, you assign the main task as a continuous task, but you can change its characteristics. Periodic tasks function as selectable time interrupts because they are triggered at a repetitive time period. This interruption may be set from 1 ms to 2000 ms, with the default time being 10 ms. Periodic tasks of the same priority execute on a time slice basis at 1-ms intervals.

3.3.2 Programs

There can be up to 32 separate programs assigned to a task, each with its own routines and program-scoped tags. Once a task is triggered, all the programs assigned to that task will execute in the order in which they are grouped. A program can be assigned to only one task in the controller.

3.3.3 Routines

A routine is defined as a set of logic instructions in a single programming language, such as ladder logic. A routine is similar to the program files in an SLC 500 or a

PLC-5. Each of the programs has a main routine, which is the first routine to be scanned when the triggered task calls the associated program. Other logic, such as a JSR instruction, then has to call the other routines.

▶ 3.4 CONTROLLOGIX DATA FILE TYPES

The Logix5550 controller uses *tags* to identify the memory location in the controller where data is stored. There are no predefined data tables, such as those in the SLC, but there are predefined data types. If you wish to group data, you make an *array,* which is a grouping of tags of similar data types.

You may create tags before the logic is entered. You may also enter tag names as the logic is entered and define them later. Or, you may use question marks [?] in place of the tag names and assign the tags later.

3.4.1 Naming Tags

Tag names follow the IEC 1131-3 identifier rules.

They must begin with an alphabetic character (A–Z or a–z) or an underscore.

They can contain only alphabetic characters, numeric characters, and underscores.

They may be 40 characters long.

They may not have consecutive or trailing underscore characters.

They are not case-sensitive.

3.4.2 Defining Tags

When defining tags, the following information is to be specified:

Name	Name the tag.
Description	(Optional) Enter a tag description. (Descriptions may be up to 120 characters in length and are not downloaded to the controller but are stored in the offline project file.)
Tag type	Enter one of the tag types: base, alias, or consumed.
Data type	Select from the list of predefined and any user-defined data types. If the tag is to be an array, specify the number of elements in up to three dimensions.
Scope	Select the scope in which to create the tag. The options are the controller scope or any one of the existing program scopes.

Display style	This defines the default display type when you are monitoring the tag in the programming software. The programming software displays the choices of available styles. These choices depend on the data type.
Produce this tag	Select whether you want to make this tag available to other controllers. Also, specify the number of other controllers that can consume this tag.

3.4.3 Tag Types

There are three different tag types: *base, alias,* and *consumed tags.*

A base tag defines the memory location where a data element is stored. An alias tag references a memory location that has been defined by another tag. An alias tag can reference a base tag or another alias tag. A consumed tag references data that comes from another controller.

3.4.4 Data Types

The controller contains a group of *predefined data types.* There are two predefined types, *basic types,* which follow the IEC 1131-3 data definitions, and *structure types,* which is created for a predefined function using the basic types.

Basic types (also known as *atomic data types*)

BOOL	1-bit boolean
SINT	1-byte integer
INT	2-byte integer
DINT	4-byte integer
REAL	4-byte floating point

Structure types

CONTROL	Control structure for array instructions
COUNTER	Control structure for counter instructions
MOTION_INSTRUCTION	Control structure for motion instructions
PID	Control structure for the PID instruction
TIMER	Control structure for timer instructions
AXIS	Control structure for an axis (controller-only tag*)
MESSAGE	Control structure for the message instruction (controller-only tag)
MOTION_GROUP	Control structure for a motion group (controller-only tag)

---

*Controller-only tags do not support arrays, cannot be passed to other routines using the JSR instruction, and cannot be nested in user-defined structures.

3.4.5 Base Tags

Base-tag memory storage depends on the data type. The controller stores all data in a minimum of 4 bytes (32 bits of data). The data types used to define base tags are BOOL, SINT, INT, DINT, and REAL.

A BOOL base tag consists of 1 bit and will store its bit in bit 0 of the 32 bits. Bits 1 through 31 are not used.

A SINT base tag consists of 8 bits and stores its value in the low byte (bits 0–7); bits 8 through 31 are not used.

An INT base tag consists of 16 bits and uses the lower 2 bytes (bits 0–15); bits 16 through 31 are not used.

DINT and REAL base tags consist of 32 bits and use all 4 bytes (bits 0–31).

3.4.6 Structures

There are three different types of structures: *predefined, module-defined,* and *user-defined* structures. Structures store a group of data, and each member of a structure can be a different data type. Some of the predefined structures, the CONTROL, COUNTER, MOTION_INSTRUCTION, PID, and TIMER, can be used in arrays and user-defined structures.

The module-defined structures are automatically created when the I/O modules are configured for the system. These structures usually contain members for data, status information, and fault information.

User-defined structures group different types of data into a single, named entity. The structure contains one or more data definitions called members. The data type for the member determines the amount of memory allocated for that memory. The data type for each member can be a basic data type, a predefined structure, a user-defined structure, a single-dimension array of an atomic data type, a single-dimension array of a predefined structure, or a single dimension of a user-defined structure.

When you create a user-defined structure, you must name the structure, and you may use an optional description. Then you must enter the following parameters for each member:

Name	Name the member.
Data type	Select the desired data type from the list. The list consists of the predefined data types and any user-defined data types.
Style	Select the desired style of the member from the list of available styles. The available styles depend on the data type.
Description	Describe the member (optional).

TABLE 3.1 MEMORY ALLOCATION FOR A USER-DEFINED STRUCTURE

User-defined structures

Name: **Test**

Member	Data type
Data_1	SINT
Data_2	REAL
Bit_0	BOOL
Bit_1	BOOL

Bit	31 24	23 16	15 8	7 0
Data allocation 1	Unused	Unused	Unused	Data_1
Data allocation 2	Data_2			
Data allocation 3	Unused	Unused	Unused	Bit 0 Bit_0 Bit 1 Bit_1

TABLE 3.2 MEMORY ALLOCATION FOR A USER-DEFINED STRUCTURE

User-defined structures

Name: **Test_1**

Member	Data type
Data_3	SINT
Data_4	REAL
User	Test

Bit 31	24 23	16 15	8 7	0
Data allocation 1	Unused	Unused	Unused	Data_3
Data allocation 2	Data_4			
Data allocation 3	Unused	Unused	Unused	Data_1
Data allocation 4	Data_2			
Data allocation 5	Unused	Unused	Unused	Bit 0 Bit_0 Bit 1 Bit_1

Memory allocation for a user-defined structure depends on the data type of the member within the instruction.

Table 3.1 shows the memory allocation for a user-defined instruction.

Table 3.2 shows a second example of a user-defined structure. In this example, one member of the user-defined structure is a user-defined structure. The member is taken from the previous example. Data allocations 3, 4, and 5 are from the user-defined structure shown in Table 3.1.

3.4.7 Members

Members within a structure are addressed by using the tag name and a period, followed by the member name (tag_name.member_name). If a structure is within another structure, the tag name of the highest level is followed by the substructure tag name, followed by the member name (tag_name.substructure_name.member_name).

3.4.8 Arrays

Arrays allow the grouping of sets of data of the same data type in a contiguous block of controller memory. A single tag within the array is one element. The element may be a basic type or a structure. The elements start with 0 and extend to the number of elements minus 1.

An array can have up to three dimensions, unless it is a member of a structure. Then it can have only one dimension. Arrays support user-defined data types and any predefined data type except Axis, Message, or Motion_Group.

Figure 3.1 illustrates the memory layout of a one-dimensional array, with the tag name of Model_Data and the data type DINT[7], which indicates an array consisting of 7 elements. The elements, as shown, are numbered 0–6.

A two-dimensional array is shown in Figure 3.2. The first column represents the style of a part and the second column represents the color of the part. The array contains 14 elements.

Figure 3.3 illustrates a three-dimensional array. One dimension represents the model number, the second dimension represents the color, and the third dimension represents the quantity to be made. There are 42 elements in the array.

In an array there are element addresses and bit addresses. Each type may be numeric (fixed) or variable. In the examples in Table 3.2 and Figures 3.1 and 3.2, numeric addressing was used. One or more numbers were placed inside the square

Array—Model_Data
Data Type—DINT[7]

	Model_Data[0]
	Model_Data[1]
	Model_Data[2]
	Model_Data[3]
	Model_Data[4]
	Model_Data[5]
	Model_Data[6]

Figure 3.1   One-dimensional array.

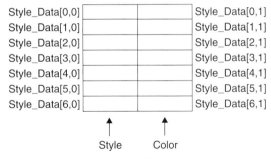

Array—Style_Data
Data Type—DINT[7,2]

Style_Data[0,0]	Style_Data[0,1]
Style_Data[1,0]	Style_Data[1,1]
Style_Data[2,0]	Style_Data[2,1]
Style_Data[3,0]	Style_Data[3,1]
Style_Data[4,0]	Style_Data[4,1]
Style_Data[5,0]	Style_Data[5,1]
Style_Data[6,0]	Style_Data[6,1]

Style   Color

Figure 3.2 Two-dimensional array.

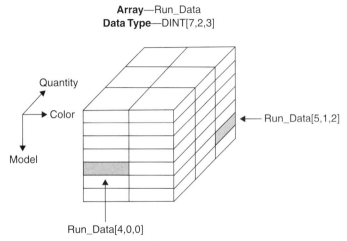

Array—Run_Data
Data Type—DINT[7,2,3]

Quantity

Color

Model

Run_Data[5,1,2]

Run_Data[4,0,0]

Figure 3.3 Three-dimensional array.

brackets, which made the addresses fixed. To make an address a variable address, the following format is used:

Array_Name[Tag or Expression]

Examples of variable element addresses are the following:

Model[Color]
Model[Quantity-4]

The mathematical operators that can be used in an expression to specify an array element are shown in Table 3.3.

TABLE 3.3 MATHEMATICAL OPERATORS

Operator	Description
+	Add
−	Subtract
*	Multiply
/	Divide
AND	And
FRD	From BCD to integer
NOT	Complement
OR	Or
TOD	From integer to BCD
SQR	Square root
XOR	Exclusive Or

When using a variable address in an array, the address must always fall within the boundaries of the array. If the dimension of the array is exceeded, a major fault is generated.

Numeric-bit addressing within the array uses the following format:

Array_Name[Element Number].Bit
Example: Model_Data[2].3

Variable-bit addressing within the array uses the following format:

Array_Name[Element Number].Tag or Expression
Example: Model[Color].Index-2

3.4.9 ControlLogix Data Storage Summary

Figure 3.4 defines tags. Tags are the basic identifiers for memory locations. They are used for allocating memory, referencing data in the logic, and monitoring data.

Tag—name & data type
- Base tag
- Alias tag
- Consumed tag
- Always allocated as DINT (32 bits)

- **Base tag**–defines memory where the data is actually stored
- **Alias tag**–references memory defined by another tag
- **Consumed tag**–data value comes from another controller

Figure 3.4 Tags.

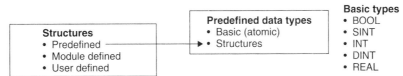

Structures
- Predefined
- Module defined
- User defined

Predefined data types
- Basic (atomic)
- Structures

Basic types
- BOOL
- SINT
- INT
- DINT
- REAL

Module defined
- Automatically created for each I/O module

User defined
- Groups different types of data into a single named entity
- An element of a structure is called a member
- Member has its own name and data type
- Member can be a structure
- Optimizes memory

Predefined structures
- AXIS
- CONTROL*
- COUNTER*
- MESSAGE
- MOTION_GROUP
- MOTION_INSTRUCTION*
- PID*
- TIMER*

*Can be used in an array or user-defined structure

Figure 3.5   Structures and predefined data types.

Array
1-, 2-, or 3-dimensional

- An array is a numerically indexed sequence of tags of the same data type that occupy a contiguous block of memory.
- Single tag in an array is an element.
- An array that is made up of a member of a structure can only have one dimension.
- Supports user-defined and predefined data types except Axis, Message, or Motion_Group.

Figure 3.6   Arrays.

Figure 3.5 shows structures and predefined data types. Structures store different data types. Predefined data types are either basic, also called atomic, or predefined structures. Structures can be used to optimize memory.

Arrays are defined in Figure 3.6. Arrays combine the same data type in a contiguous block of memory.

SUMMARY

The SLC 500 and the ControlLogix controllers use different methods of assigning memory for projects. The SLC 500 divides memory into two basic areas, program files and data files. Program files contain controller information, the main ladder program, and any subroutine programs. The main program is stored in program file 2, and program files 3–255 may be used to store subroutines. The SLC 500 assigns nine default data files, and files 9–255 may be assigned. There is a structure assigned to each data file, such as timer, counter, or integer.

A project file that contains programming and configuration information is created for a ControlLogix controller. The project will have three major components: tasks, programs, and routines. Tasks schedule and provide priority for how programs or a group of programs assigned to a task execute. Routines make up a program and consist of a set of logic instructions in a single programming language, such as ladder logic. The routines are similar to the program files in an SLC 500. Memory locations in the data-storage area of memory are identified by tags. There are no predefined data tables, as in the SLC 500, but there are predefined data types. A group of similar data types is referred to as an array.

▶ ## EXERCISES

1. What is the maximum number of data files and program files in an SLC 500?
2. Where is the main program file located in an SLC 500?
3. What is stored in program files 3–255 in an SLC 500?
4. What are the default data files in an SLC 500?
5. In an SLC 500 system, what is the address of an input terminal number 5 located on a module located in slot 3?
6. In an SLC 500 system, what is the address of output terminal number 8 located on a module located in slot 7?
7. In an SLC 500, what is the address of the preset value of timer 6 located in data file 22?
8. In an SLC 500, what is the address of the done bit of counter 8 located in data file 30?
9. In an SLC 500, in what data type would a whole number within the range −32,768 through +32,767 be stored?
10. In an SLC 500, in what data file type would a fractional value be stored?
11. In an SLC 500, data files 9–255 are what type?
12. In a ControlLogix controller, what are the three major components of a project and what are their functions?
13. In a ControlLogix controller, what are the types of tasks and how many can there be?
14. In a ControlLogix controller, how many programs may be assigned to a task? Can a program be assigned to more than one task?
15. In a ControlLogix controller, what is the location called where the ladder logic is stored?

16. In a ControlLogix controller, what name is given to the memory location where data are stored?

17. In a ControlLogix controller, what are the three different tag types called?

18. In a ControlLogix controller, list the basic type data types and the types of data they store.

19. In the ControlLogix controller, list the predefined structure-type data types and the types of data they store.

20. In the ControlLogix controller, list the three different types of structures.

21. In a ControlLogix controller, given the following structure, lay out the memory allocated for the structure.

Name: Exercise

Member	Data Type
Data_1	REAL
Data_2	REAL
Data_3	SINT
Data_4	SINT
Bit_0	BOOL
Bit_1	BOOL

22. In a ControlLogix controller, arrays group data of what types?

23. In a ControlLogix controller, what is the maximum number of dimensions in an array?

24. In a ControlLogix controller, how many elements are in an array with the tag DATA and the data type DINT[8]?

25. In a ControlLogix controller, how many elements are in an array with the tag MODEL and the data type DINT[6,4,5]?

4 SLC 500 and ControlLogix Bit Instructions

CHAPTER OUTLINE

LEARNING OBJECTIVES

Upon reading this chapter the students should be able to:

- Apply addressing to bit-level instructions.
- Apply Examine If Closed (XIC) instructions.
- Apply Examine If Open (XIO) instructions.
- Apply Output Energize (OTE) instructions.
- Recognize the differences between XIO and XIC instructions and relay contacts.
- Apply Output Latch (OTL) and Output Unlatch (OTU) instructions.

INTRODUCTION

This chapter identifies the bit-level instructions used in ladder logic. The instructions include the **input instructions** examine on (XIC) and examine off (XIO) as well as the output instructions output energize (OTE), output latch (OTL), and output unlatch (OTU). Examples of their use in the ladder logic project are also given.

▶ 4.1 ADDRESSING BIT-LEVEL INSTRUCTIONS

Addressing bit-level instructions requires bit-level (Boolean) addresses. The format of the bit-level addresses depends on whether you are using an SLC 500 or a ControlLogix processor.

4.1.1 SLC 500 Bit-Level Addressing

Addressing in the SLC 500 is done at three different levels: the element level, the word level, and the bit level. Some data table types have multiple words per element, such as the timer type, which has three words per element. Each element consists of a control word, a preset word, and an accumulated word. Each individual word in an element address is referred to as a subelement. Integer elements have one word per element. Each word has 16 bits, which are numbered from 0 to 15. When addressing to the bit level, the address always refers to the bit within the word, or subelement. You cannot address individual bits in the floating-point data type.

The format for a bit-level address in the SLC 500 is

file type file number:word/bit

The following examples illustrate the different types:

Integer bit N7:10/15
Binary bit B3:1/5 or B3/21 (depending on the display format)
Input bit I:3/5
Timer accumulated bit T4:0.ACC/2

4.1.2 ControlLogix Bit-Level Addressing

In a ControlLogix processor, bits can be referenced in a base tag, a member within a structure, or a bit within an element of an array.

In a base tag, the bit is referenced by the tag name of a BOOL data type—for example, *pump_1*. In a structure, the bit is referenced by the tag name of the BOOL data type member—for example, *counter_1.dn*. In an array, the format for a bit address is ArrayName[Element Number].[Tag or Expression]. An example of a numeric bit is *motor[1].5*, and an example of a variable bit is *motor[1].[index+5]*.

▶ 4.2 INSTRUCTIONS

4.2.1 Examine If Closed (XIC) Instruction, -] [-

The examine if closed (XIC) instruction is an input instruction that examines or reads the status of a single bit at the bit location determined by the address on the instruction to see if it is 0 or 1.

If the examined bit is a 1, the instruction is logically true, as shown in Figure 4.1.

If the examined bit is a 0, the instruction, as shown in Figure 4.2, is logically false.

4.2.2 Examine If Open (XIO) Instruction, -]/[-

The examine if open (XIO) instruction is an input instruction that examines a single bit at the bit location determined by the address on the instruction to see if it is 0 or 1.

If the examined bit is a 0, the instruction, as shown in Figure 4.3, is logically true.

If the examined bit is a 1, the instruction, as shown in Figure 4.4, is logically false.

Figure 4.1   XIC instruction logically true.

Figure 4.2   XIC instruction logically false.

Figure 4.3   XIO instruction logically true.

Figure 4.4   XIO instruction logically false.

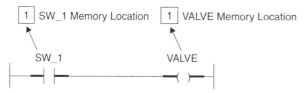

Figure 4.5 OTE rung logically true.

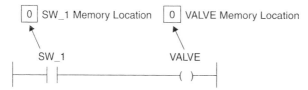

Figure 4.6 Rung logically false.

4.2.3 Output Energize (OTE) Instruction, -( )-

The output energize (OTE) instruction is an output instruction that writes to the data table address on the instruction when it is true or false. Whether it is true or false depends on the continuity of the rung logic on which the output energize instruction is located.

Figure 4.5 illustrates an output energize instruction on a rung with true logic. If SW_1 is closed, there is a 1 at the SW_1 memory location. The examine if closed instruction is true, making the rung true. The output energize instruction then writes a 1 to the VALVE memory location.

The next rung, in Figure 4.6, illustrates an output instruction on a rung with false logic. If SW_1 is open, there is a 0 at the SW_1 memory location. The examine if closed instruction is false, resulting in a false rung. The output energize instruction then writes a 0 to the VALVE memory location.

▶ 4.3 COMPARING XIC AND XIO TO NORMALLY OPEN AND NORMALLY CLOSED RELAY CONTACTS

The symbols used for the XIC and XIO instructions go back to the electrical diagrams used with hardwired logic systems, which used relays for logic control. The symbol -] [- used for the XIC instruction is a normally open symbol in a relay logic circuit. The symbol -]/[- used for the XIO instruction is a normally closed symbol in a relay logic circuit. This has caused confusion in programmable controller logic diagrams, as people often refer to the instructions as normally open and normally

SLC 500 input module located in slot 3

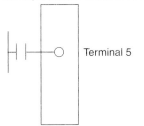

Terminal 5

Normally open relay contact wired to terminal 5

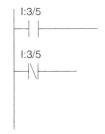

I:3/5

I:3/5

Normally open contact's status
may be examined with either
the XIC or XIO instruction

Figure 4.7   Normally open contact.

closed contacts. Figure 4.7 illustrates the difference between a normally open contact used with the examine on and examine off instructions.

Even though they are represented by the same symbol, the relay contact on the left is a hardwired device, whereas the XIC on the right side of the diagram is an instruction in the processor's ladder-logic program.

▶ 4.4 OUTPUT LATCH (OTL), -(L)-, AND OUTPUT UNLATCH (OUT), -(U)-, INSTRUCTIONS

The output latch (OTL) and the output unlatch (OTU) instructions are retentive *output instructions* in that they take action only when the rung they are on goes true. They do not take action when the rung goes false. The OTL and OTU instructions are usually used together with the same address, but they may also be used independently of each other.

The output latch (OTL) instruction sets a bit address to 1 when the instruction goes true and does nothing when the rung goes false. The bit remains 1.

The OTU instruction resets a bit to 0 when the instruction goes true and does nothing when the rung goes false. The bit remains 0.

The following example shows the latch and unlatch instructions using the same address. In Figure 4.8, the first rung, with the latch instruction, is true. The second rung, with the unlatch, is not. The INTERNAL bit in memory will be set. The purpose of the XIO instructions on each rung is to keep the latch and unlatch instructions from going true at the same time.

In Figure 4.9, when SW_1 is turned off, the INTERNAL bit remains set. When the latch goes false, it does not take any action.

When SW_2 is turned on, as shown in Figure 4.10, the INTERNAL bit is reset, and neither the OTL nor the OTU are highlighted.

In Figure 4.11, SW_2 is turned off and the INTERNAL bit remains reset.

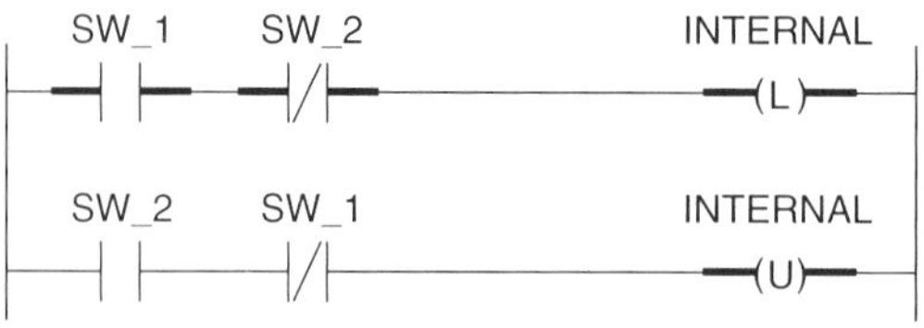

Figure 4.8 Latch and unlatch instructions.

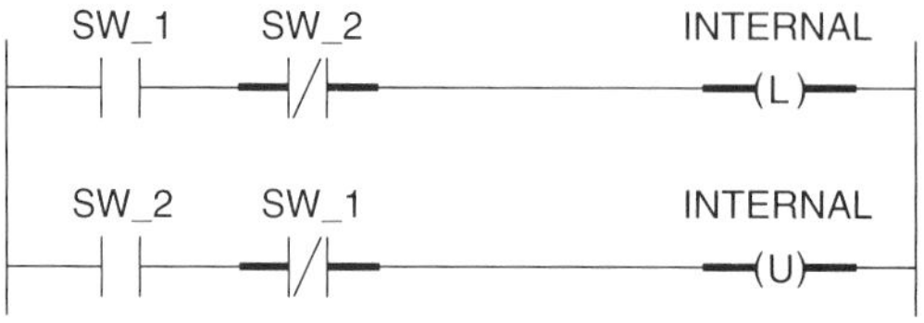

Figure 4.9 Latch and unlatch rung false.

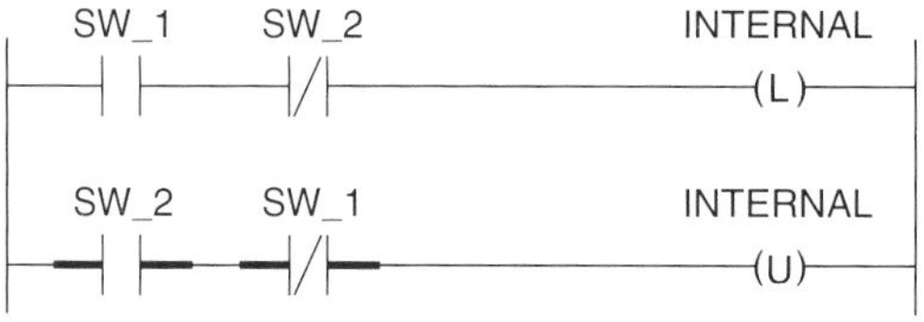

Figure 4.10 Unlatch rung true.

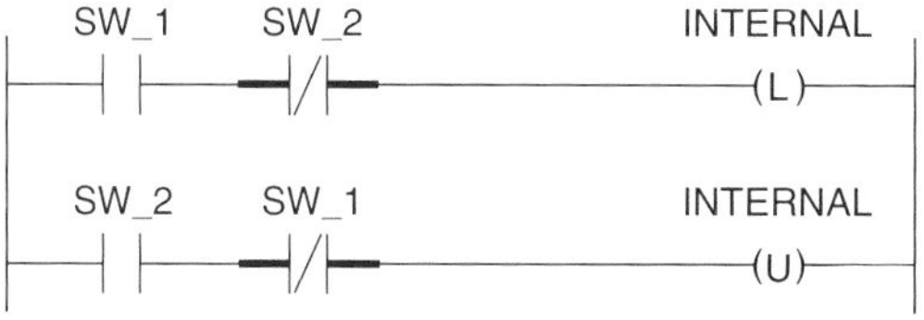

Figure 4.11 Latch and unlatch rungs false.

SUMMARY

Bit-level instructions examine the status of a bit and control the status of bits. The XIC and the XIO are input instructions that examine the status of bits. The XIC is true if it finds the status of the bit at its bit address is 1. The XIO is true if it finds the status of the bit at its bit address is 0. The OTE instruction is an output instruction that controls the status of the bit at its bit address. When the OTE is logically true, it sets the bit, and if it is logically false, it resets the bit. The OTL and OTU are retentive output instructions that influence the status of a bit only when they are true. The OTL sets a bit, and the OTU resets a bit.

The symbols used for the XIC and XIO instructions are the same as those used for normally open and normally closed instructions in a hardwired relay diagram. This causes confusion for those who have worked with hardwired systems and now have to interpret these symbols differently in a programmable controller logic diagram.

▶ EXERCISES

1. What is the symbol for an XIC instruction and what determines whether the instruction is logically true or false?
2. What is the symbol for an XIO instruction and what determines whether the instruction is logically true or false?
3. How does the OTE instruction control the status of the bit at its bit address?
4. Design the ladder logic for turning on an output when a normally closed switch (SW_1) is closed and a normally open switch (SW_2) is closed. When SW_1 is open or when SW_2 is open, the output will be off.
5. Design the ladder logic for turning on an output when a normally closed switch (SW_3) is open or a normally closed switch (SW_4) is open.
6. In Exercises 4 and 5, will the ladder logic be different if the status of the switches necessary to turn on the outputs remained the same but the normal position of the switches was different?
7. Design the ladder logic for turning on an output when a switch (SW_5) is momentarily closed and then opened. When SW_5 opens, the output remains on.
8. Add logic to Exercise 7 so that when a switch (SW_6) is momentarily closed, the output will turn off. When SW_6 is turned off, the output will remain off unless SW_5 is on. The logic will prevent the rung in Exercise 7 from going true when SW_6 is turned on and also prevent the rung in Exercise 8 from going true when SW_5 is on.
9. Design the ladder logic that will turn on an output when a switch (SW_7) is closed and turn on a second output when SW_7 is open. The two outputs will not be on at the same time.
10. Design the ladder logic that will turn on an output when SW_8 is closed or SW_9 is open and SW_10 is closed.

5 SLC 500 and ControlLogix Timer and Counter Instructions

CHAPTER OUTLINE

LEARNING OBJECTIVES

Upon reading this chapter the students should be able to:

- Apply on-delay timers.
- Apply off-delay timers.
- Apply retentive timers.
- Apply count up counters.
- Apply count down counters.

INTRODUCTION

This chapter identifies the timer and counter instructions used in ladder logic. The timer instructions include the timer on delay (TON), the timer off delay (TOF), and the retentive timer (RTO). The counter instructions include the count up counter (CTU) and the count down counter (CTD). Examples are also given that show how they are used in the ladder-logic project.

▶ 5.1 TIMERS

Timers are used to turn outputs on or off after a time delay, turn outputs on or off for a set amount of time, or keep track of time an output is on or off. The SLC 500 and ControlLogix function in the same manner. The differences occur in the time base, timer address, and the maximum values of the preset and accumulated values. The timer address in the SLC 500 is a data table address or symbol, as explained in Chapter 3, whereas the timer address in the ControlLogix controller is a predefined structure of the TIMER data type (also explained in Chapter 3). In the SLC 500, the maximum value for the preset and accumulated values is 32,767, and in the ControlLogix processor, the maximum value is 2,147,483,647.

The timer structure for all timers is as follows.

Mnemonic	Description
en	Enable bit: the bit set when the instruction is enabled.
tt	Timer timing bit: the bit set when the timer is timing.
dn	Done bit: the bit set when the accumulated value equals the preset value.
.pre	Preset value in time-base multiples: the accumulated value must reach this value before the DN bit is set.
.acc	Accumulated value in time-base multiples: this time represents the elapsed time after the timer is enabled.
Time Base	In the SLC 500, there is a choice of time base, either 1.0 s or 0.01 s. In the ControlLogix, the time base is fixed at 1 ms.

5.1.1 Timer On Delay (TON)

The timer on delay (TON) instruction is a nonretentive output instruction used to turn outputs on or off after a time delay. The TON timer starts timing when it goes true and times until the accumulated value reaches the preset value or until it goes false. If it has timed out—that is, the accumulated value equals the preset value—the dn bit will be set. The timer will reset whenever it goes false.

Figures 5.1 and 5.2 show the TON timer instructions for the SLC 500 and the ControlLogix controller. The difference between them is that the ControlLogix timer has a fixed time base.

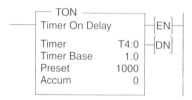

Figure 5.1   SLC 500 TON timer.

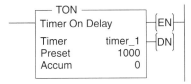

Figure 5.2 ControlLogix TON timer.

```
  sw_1                           ___ TON _____
───┤ ├────────────────────────┤ Timer On Delay        ─(EN)─
                               │                            
                               │ Timer       timer_1   ─(DN)
                               │ Preset        10000        
                               │ Accum             0        
                               └────────────────────────────


  timer_1.en                                    lt_0
───┤ ├────────────────────────────────────────( )──────


  timer_1.tt                                    lt_1
───┤ ├────────────────────────────────────────( )──────


  timer_1.dn                                    lt_2
───┤ ├────────────────────────────────────────( )──────
```

Figure 5.3 TON example.

Figure 5.3 shows a TON timer with its status bits controlling outputs on the rungs following the timer. The timer shown represents a ControlLogix display, but the SLC 500 timer functions in the same way.

The status bit operation is as follows.

en bit   Follows the status of the timer. Will be set whenever sw_1 is on.

tt bit   Set when the timer is timing. When sw_1 is turned on and remains on for more than 10 s, the tt bit is set for 10 s. lt_1 is set whenever the tt bit is set.

dn bit   Set when the accumulated value is equal to the preset value. lt_2 is on when the dn bit is set. The dn bit may be used to turn on an output time delay after an input initiates the timer.

5.1.2 Timer Off Delay (TOF)

The timer off delay (TOF) instruction is a nonretentive output instruction used to turn outputs on or off after a time delay. The TOF timer starts timing when it goes false and times until the accumulated value reaches the preset value or until it goes true and resets. If it has timed out—that is, the accumulated value equals the pre-set value—the dn bit is reset. The timer resets when it goes true.

Figures 5.4 and 5.5 show the TOF timer instructions for the SLC 500 and the ControlLogix controller. The difference is the fixed time base in the ControlLogix timer.

Figure 5.6 shows a TOF timer with its status bits controlling outputs on the rungs following the timer. The timer shown represents an SLC 500 display, but the ControlLogix timer functions in the same way.

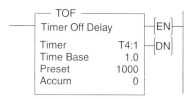

Figure 5.4   SLC 500 TOF timer.

Figure 5.5   ControlLogix TOF timer.

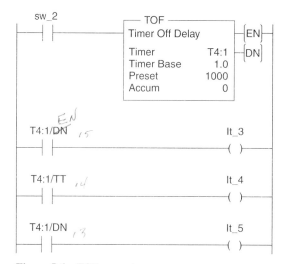

Figure 5.6   TOF example.

The status bit operation is as follows.

en bit   Follows the status of the timer. This bit is set whenever sw_2 is on.

tt bit   Set when the timer is timing. When sw_2 is turned off and remains off for more than 10 s, the tt bit is set for 10 s. lt_4 is set whenever the tt bit is set.

dn bit   Set when the timer goes true. The bit resets when the timer goes false and the accumulated value is equal to the preset value. lt_5 is on when the dn bit is set.

5.1.3 Retentive Timer On Delay (RTO)

The retentive timer on delay (RTO) instruction is a retentive output instruction used to turn outputs on or off after a time delay. The RTO timer starts timing when it goes true and times until the accumulated value reaches the preset value or it goes false. It retains its accumulated value when it goes false. When the timer goes true again, it resumes timing from that accumulated value. When it has timed out—that is, the accumulated value equals the preset value—the dn bit is set. The timer is reset by the reset (RES) instruction.

Figures 5.7 and 5.8 show the RTO timer instructions for the SLC 500 and the ControlLogix controller. The difference is the fixed time base in the ControlLogix timer.

Figure 5.9 shows an RTO timer with its status bits controlling outputs on the rungs following the timer. The timer shown represents a ControlLogix display, but the SLC 500 timer functions in the same way.

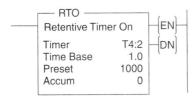

Figure 5.7 SLC 500 RTO timer.

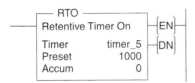

Figure 5.8 ControlLogix RTO timer.

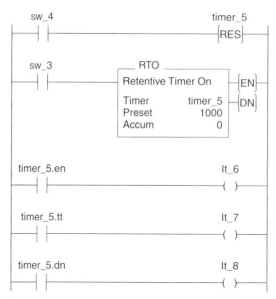

Figure 5.9 RTO example.

The RTO instruction requires an RES instruction to reset the timer. The RES instruction has the same address as the timer it is to reset. When the RES instruction is true, the accumulated value resets to zero and the status bits is reset.

The status bit operation is as follows.

en bit   Follows the status of the timer. This bit is set whenever sw_3 is on.

tt bit   Set when the timer is timing. The timer is timing when sw_3 is on and the accumulated value has not reached the preset value. lt_1 is set whenever the tt bit is set.

dn bit   Set when the accumulated value is equal to the preset value, even when the timer instruction goes false. lt_2 is on when the dn bit is set. The dn bit may be used to turn on an output time delay after an input initiates the timer.

► 5.2 COUNTERS

Counters are used to turn outputs on or off after a certain number of input transitions. There are two counter instructions, the count up counter (CTU) and the count down counter (CTD). The reset (RES) instruction is used to reset the accumulated value and the status bits of the counter. The SLC 500 and ControlLogix function in the same manner. The counter address and the maximum values of the preset and accumulated values differ. The counter address in the SLC 500 is a data table address or symbol, as explained in Chapter 3, whereas the timer address in the ControlLogix controller is a predefined structure of the COUNTER data type (also explained in Chapter 3). In the SLC 500, the maximum value for the preset and accumulated values is 32,767 and the minimum value is −32,768; in the ControlLogix processor, the maximum value is 2,147,483,647 and the minimum value is −2,147,483,648.

The counter structure for both counters is as follows.

Mnemonic	Description
cu	Count up counter enable bit: the bit set when the count up counter is enabled.
cd	Count down counter enable bit: the bit set when the count down counter is enabled.
dn	Done bit: the bit set when the accumulated value equals the preset value.
ov	Overflow bit: the bit set when the count up counter counts past the maximum value (SLC 500, 32,767; and ControlLogix, 2,147,483,647). The counter then goes to −32,768 in the SLC 500 or −2,147,483,648 in the ControlLogix controller and starts counting up again.
un	Underflow bit: the bit set when the count down counter counts past the lower-limit value (SLC 500, −32,768; and ControlLogix, −2,147,483,648). The counter then goes to 32,767 in the SLC 500 or 2,147,483,647 in the ControlLogix controller and starts counting down again.
.pre	Preset value: the value the accumulated value must reach before the dn bit is set.
.acc	Accumulated value: this value represents the number of transitions of the counter instruction.

5.2.1 Count Up Counter (CTU)

The count up counter (CTU) instruction is a retentive output instruction used to turn outputs on or off after a certain number of transitions of the instruction. The CTU instruction increments its accumulated value by 1 every time it is transitioned. It sets its done bit when the accumulated value is equal to or greater than the preset value. The CTU instruction requires the RES instruction to reset its accumulated value and its status bits.

Figures 5.10 and 5.11 illustrate the CTU counters for the SLC 500 and ControlLogix controllers. The CTU counters for the SLC 500 and the ControlLogix controllers are identical except for the upper limits of the preset and accumulated values.

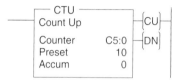

Figure 5.10   SLC 500 CTU counter.

```
        ┌─ CTU ──────────────┐
        │ Count Up           │──(CU)──
        │                    │
        │ Counter  counter_1 │──(DN)──
        │ Preset         100 │
        │ Accum            0 │
        └────────────────────┘
```

Figure 5.11   ControlLogix CTU counter.

```
   sw_5                                counter_1
  ──┤ ├──────────────────────────────────(RES)──

   sw_6              ┌─ CTU ──────────────┐
  ──┤ ├──────────────│ Count Up           │──(CU)──
                     │                    │
                     │ Counter  counter_1 │──(DN)──
                     │ Preset         100 │
                     │ Accum            0 │
                     └────────────────────┘

   counter_1.cu                         lt_9
  ──┤ ├──────────────────────────────────( )──

   counter_1.dn                         lt_10
  ──┤ ├──────────────────────────────────( )──

   counter_1.ov                         lt_11
  ──┤ ├──────────────────────────────────( )──
```

Figure 5.12   CTU example.

Figure 5.12 shows a CTU counter with its status bits controlling outputs on the rungs following the timer. The counter illustrated represents a ControlLogix display, but the SLC 500 counter functions in the same way.

The CTU instruction requires an RES instruction to reset the counter. The RES instruction has the same address as the counter it is to reset. When the RES instruction is true, the accumulated value resets to zero and the status bits are reset.

The status bit operation is as follows.

cu bit   Follows the status of the counter. It is set whenever sw_6 is on.

dn bit   Set when the accumulated value is equal to or greater than the preset value. lt_10 will be on when the dn bit is set.

ov bit   Set when the counter has exceeded its upper limit.

5.2.2 Count Down Counter (CTD)

The count down counter (CTD) instruction is a retentive output instruction used to turn outputs on or off after a certain number of transitions of the instruction. The CTD instruction decrements its accumulated value by 1 every time it is transitioned. It sets its done bit when the accumulated value is equal to or greater than the preset value. The CTD instruction requires the RES instruction to reset its accumulated value and its status bits. It resets its accumulated value to 0. The CTD instruction then counts negative when it transitions.

Figures 5.13 and 5.14 illustrate the CTD counters for the SLC 500 and ControlLogix controllers. The CTD counters for the SLC 500 and the ControlLogix controllers are the same except for the upper limits of the preset and accumulated values.

Figure 5.15 shows a CTU counter and a CTD counter used together with the same address. This is the most common application of the CTD counter. It counts negative, so if it were used by itself with a preset, positive value, its dn bit would be reset when its accumulated value was zero. Then, counting in a negative direction, it would never reach its preset value to set the dn bit. However, the preset can be entered with a negative value; then the dn bit is set until the accumulated value becomes less than the preset value.

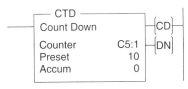

Figure 5.13   SLC 500 CTD counter.

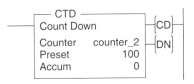

Figure 5.14   ControlLogix CTD counter.

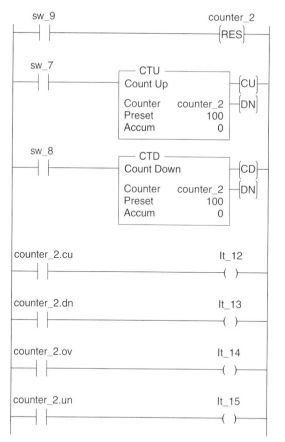

Figure 5.15 CTD example.

As shown in Figure 5.15, the counter pair makes up what is called an up-down counter. Every time the CTU initializes, 1 is added to the accumulated value of both counters, because they have the same address. Every time the CTD initializes, 1 is subtracted from the accumulated value. This combination can be used to keep track of the number of parts in a holding area. Every time a part enters, it turns on sw_7, increasing the count by 1. Every time a part leaves, it turns on sw_8, decreasing the count by 1. The accumulated value represents the number of parts in the holding area. The dn bit could be used, when set, to stop any more parts from entering the holding area.

Figure 5.15 shows the counters with their status bits controlling outputs on the rungs following the counter. The counters illustrated represent a ControlLogix display, but SLC 500 counters function in the same way.

The CTD instruction requires an RES instruction to reset the counter. The RES instruction has the same address as the counter it is to reset. When the RES instruction is true, the accumulated value resets to 0 and the status bits are reset. In the example, the RES resets both the CTU and the CTD instructions.

Status bit operation is as follows:

cu bit — Follows the status of the up counter. This bit is set whenever sw_7 is on.

cd bit — Follows the status of the down counter. This bit is set whenever sw_8 is on.

dn bit — Set when the accumulated value is equal to or greater than the preset value. lt_13 is on when the dn bit is set.

ov bit — Set when the counter has exceeded its upper limit.

un bit — Set when the counter has exceeded its lower limit.

SUMMARY

Timers are used to turn outputs on or off after a time delay. There are three different timers: timer on delay (TON), timer off delay (TOF), and retentive timer (RTO).

The TON starts timing when it goes true and resets when it goes false. Its done bit is set when the accumulated value is equal to the preset value.

The TOF starts timing when it goes false and resets when it goes true. Its done bit is set when the TOF resets, and the done bit resets when the accumulated value reaches the preset value.

The RTO timer is an on-delay timer that is retentive. When it goes false, it retains its accumulated value and resumes timing from that point when it goes true until it reaches its preset value. The reset (RES) instruction is required to reset the RTO instruction.

Counters count transitions of the counter-enable bit from false to true. They are used to count transitions of inputs and then control outputs after a certain number of counts. There are two counter instructions, the count up counter (CTU) and the count down counter (CTD). The CTU increases its accumulated value by 1 each time it sees a transition, and the CTD decreases its accumulated value by 1 each time it sees a transition. The CTD is frequently used in conjunction with a CTU to form an up-down counter function.

▶ ## EXERCISES

1. What timer(s) reset when they go false?
2. What are the timer bases for SLC 500 and ControlLogix timers?
3. What are the maximum preset values for the SLC 500 and ControlLogix timers?

4. Design the ladder logic that will turn on an output 5 s after a switch (SW_1) is turned on.

5. Design the ladder logic that will turn on an output when a switch (SW_2) is turned on. The output will turn off 7 s after SW_2 is turned off.

6. A switch (SW_3) cycles on and off. Design the ladder logic that will time the number of seconds that SW_3 is on up to a maximum of 30 s. A second switch (SW_4) will reset the accumulated value back to 0.

7. Design the ladder logic that will turn on an output (MOTOR_1) 5 s after a switch (SW_5) is turned on. A second output (MOTOR_2) will turn on 7 s after MOTOR_1 is turned on. A third output (MOTOR_3) will turn on 10 s after MOTOR_2 is turned on. All three outputs will turn off when SW_5 is turned off.

8. What is the range of counts for the SLC 500 and ControlLogix counters?

9. Design the ladder logic that will turn on an output (SOL_1) when a switch (SW_8) has gone from off to on 50 times. A second switch (SW_9) will reset the counter.

6 SLC 500 and ControlLogix Data Collecting

CHAPTER OUTLINE

6.1   Word-Level Data-Manipulation Instructions
6.2   Compute and Math Instructions
6.3   Comparison Instructions
6.4   Logical Instructions

LEARNING OBJECTIVES

Upon reading this chapter the students should be able to:

- Interpret and program word-level data-manipulation instructions.
- Interpret and program word-level arithmetic instructions.
- Interpret and program word-level comparison instructions.
- Interpret and program word-level logical instructions

INTRODUCTION

This chapter defines the structure and operation of various word-level instructions in the SLC 500 and ControlLogix controllers. There is very little difference between the controllers in the function of the instructions. Any differences are pointed out where they occur.

Word-level instructions include data-manipulation instructions, arithmetic instructions, comparison instructions, and logical instructions. These instructions allow the programmer to construct logic which will move data, a word or selected bits, from one memory location to another, do math on data, compare data between two memory locations, and do logical operations between two words of data.

▶ 6.1 WORD-LEVEL DATA-MANIPULATION INSTRUCTIONS

Data-manipulation instructions copy data between two memory locations. In the SLC 500, depending on the instruction, they may copy either a word of data from one memory location to another, or they may selectively copy anywhere from a single bit up to 16 bits between two memory locations. In ControlLogix, depending on the instruction, the data type used affects the amount of data copied.

6.1.1 Move (MOV) Instruction

The move (MOV) instruction is an output instruction that copies data from the source to the destination. When the instruction is true, it executes every time it is scanned. The source value remains unchanged when the instruction executes.

The instruction is the same for both the SLC 500 and ControlLogix. In the example in Figure 6.1, the MOV instruction copies the source to the destination.

SLC 500.

The source is a word address or a constant

The destination is a word address

Floating point may also be used for the source, destination, or both. If floating point is used for the source and has a value other than an integer value between –32,768 and 32,767 and the destination address is a word address, it will set the math overflow bit.

ControlLogix.

The source is SINT, INT, DINT or REAL.

The destination is SINT, INT, DINT or REAL.

SINT or INT converts to a DINT value by sign-extension.[*]

Figure 6.1   MOV instruction.

---

[*]Sign-extension converts data by placing the value of the leftmost bit (the sign of the value, 0=positive or 1=negative) into each bit position to the left of the existing bits until there are 32 bits.

DINT and REAL are the optimal data types for both the source and the destination. If data types are mixed and tags are used that are not the optimal type, the processor converts the data. Data conversion takes additional time and memory.

6.1.2 Masked Move (MVM) Instruction

The masked move (MVM) instruction is an output instruction that copies data through a *mask* from the source to the destination. The mask allows only a portion of the data to be copied. A 1 in the mask allows the source bit to be copied to the destination bit. A 0 in the mask prevents the source bit from being copied to the destination bit, and thus the destination bit remains in its last state.

Figure 6.2 illustrates the MVM instruction copying data from the source to the destination with a hexadecimal mask being displayed.

Figure 6.3 shows the mask displayed in the binary equivalent of the hexadecimal mask given in the MVM instruction. The shaded bits are copied from the source to the destination. The unshaded bits in the destination are in the same state as they were before the instruction executed. The source always remains unchanged in the execution of the MVM instruction.

SLC 500.

The source is any 16-bit word address, either element or subelement. Floating-point data types or other multiword elements are not permitted.

The mask may be entered as a binary value, hexadecimal value, decimal value, or word-level address. Regardless of how it is entered, the mask value is displayed in hexadecimal.

The destination is any 16-bit word address, element, or subelement. Floating-point data types or other multiword elements are not permitted.

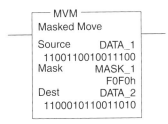

```
┌── MVM ──────────────┐
│ Masked Move         │
│                     │
│ Source      DATA_1  │
│   1100110010011100  │
│ Mask        MASK_1  │
│             F0F0h   │
│ Dest        DATA_2  │
│   1100010110011010  │
└─────────────────────┘
```

Figure 6.2 MVM instruction.

1100	1100	1001	1100	Source
1111	0000	1111	0000	Mask
1100	0101	1001	1110	Destination

Figure 6.3 MVM instruction operation.

ControlLogix.

The source is SINT, INT, or DINT. Zero fill converts the SINT or INT tag to a DINT value.[*]

The mask is SINT, INT, or DINT. Zero fill converts the SINT or INT tag to a DINT value. When you enter the mask, the programming software defaults to decimal values. If you wish to enter the mask as hexadecimal, you must precede the value with 16#. To enter the mask as an octal number, you must precede the value with 8#, and to enter it as a binary number, precede the value with 2#.

The destination is SINT, INT, or DINT. DINT is the optimal data type for the source, mask, and destination.

▶ 6.2 COMPUTE AND MATH INSTRUCTIONS

The compute and math instructions do math operations using an expression or a specific math function. The compute instruction uses an expression to perform the math operation. The other instructions do the specific math operations defined by the instructions.

6.2.1 Compute (CPT) Instruction

The compute (CPT) instruction performs math operations, as defined in the expression. The expression may contain multiple math operations. The result of the expression is placed in the destination. The CPT instruction takes longer to execute and takes more memory than the sum of the individual math instructions required to do the same math operation. The advantage of the CPT instruction is that it allows you to enter the expression in one instruction.

In Figure 6.4 the expression is evaluating the square root of DATA1 cubed and adding the result to DATA2. The result of the expression is stored at the destination address, RESULT_1.

Figure 6.4 CPT instruction.

_____

[*]Zero fill is the process of filling in zeros to the left of the existing bits until there are 32 bits.

SLC 500.

The destination is where the result of the math operation of the expression is stored and may be a word address or a floating-point address.

The expression may be 28 characters per line, with a maximum of 255 characters. The operators in the expression may be +, −, *, | (DIV), SQR, − (NEG), AND, NOT, OR, XOR, TOD, FRD, LN, TAN, ABS, DEG, RAD, SIN, COS, ATN, ASN, ACS, LOG, or **(XPY).

ControlLogix.

The destination is where the result of the math operation of the expression is stored and may be data type SINT, INT, DINT, or REAL.

There is no limit to the length of the expression. The data type for the expression may be SINT, INT, DINT, or REAL. A SINT or INT tag converts to a DINT value by sign extension. The operators in an expression may be +, −, *, /, **, ABS, ACS, AND, ASN, ATN, COS, DEG, GRD, LN, LOG, MOD, NOT, OR, RAD, SIN, SQR, TAN, TOD, TRN, or XOR. Each operator in an expression must have one or two operands (tags or immediate values). An example is

operator(operand) or operand_1 operator operand_2

6.2.2 Add (ADD) Instruction

The add (ADD) instruction is an output instruction that adds two values and stores the result in the destination address.

In Figure 6.5 the value stored at the address of source A is added to the value stored in source B, and the result is stored in the destination address.

SLC 500.

Source A may be a word address, a floating-point address, or a constant. Either source A or source B may be a constant, but both cannot be constants.

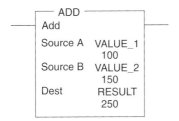

Figure 6.5 ADD instruction.

Source B may be a word address, a floating-point address, or a constant.

The destination may be a word address or a floating-point address that stores the result of the add operation.

ControlLogix.

Source A may be a SINT, INT, DINT, or REAL value, which is added to source B.

Source B may be a SINT, INT, DINT, or REAL value, which is added to source A.

The destination is a SINT, INT, DINT, or REAL tag address that stores the result.

DINT and REAL are the optimal data types.

6.2.3 Subtract (SUB) Instruction

The Subtract (SUB) instruction is an output instruction that subtracts one value from another and stores the result in the destination address.

In Figure 6.6 the value stored at the address of source B is subtracted from the value stored in source A; the result is stored in the destination address.

SLC 500.

Source A may be a word address, a floating-point address, or a constant. Either source A or source B may be a constant, but both cannot be constants.

Source B may be a word address, a floating-point address, or a constant.

The destination is a word address or a floating-point address that stores the result of the subtract operation.

ControlLogix.

Source A may be a SINT, INT, DINT, or REAL value, from which source B is subtracted.

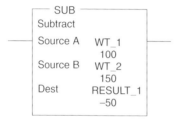

Figure 6.6 SUB instruction.

Source B may be a SINT, INT, DINT, or REAL value, which is subtracted from source A.

The destination may be a SINT, INT, DINT, or REAL tag address that stores the result.

DINT and REAL are the optimal data types.

6.2.4 Multiply (MUL) Instruction

The multiply (MUL) instruction is an output instruction that multiplies two values and stores the result in the destination address.

In Figure 6.7 the value stored at the address of source B is multiplied by the value stored in source A, and the result is stored in the destination address.

SLC 500.

Source A may be a word address, a floating-point address, or a constant. Either source A or source B may be a constant, but both cannot be constants.

Source B may be a word address, a floating-point address, or a constant.

The destination is a word address or a floating-point address that stores the result of the multiply operation.

ControlLogix.

Source A may be a SINT, INT, DINT, or REAL value, by which source B is multiplied.

Source B may be a SINT, INT, DINT, or REAL value, by which source A is multiplied.

The destination is a SINT, INT, DINT, or REAL tag address that stores the result.

DINT and REAL are the optimal data types.

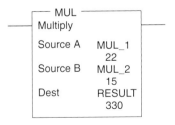

Figure 6.7 MUL instruction.

6.2.5 Divide (DIV) Instruction

The divide (DIV) instruction is an output instruction that divides source A by source B and stores the result in the destination address.

In Figure 6.8 the value stored at the address of source A is divided by the value stored in source B, and the result is stored in the destination address.

SLC 500.

Source A may be a word address, a floating-point address, or a constant. Either source A or source B may be a constant, but both cannot be constants.

Source B may be a word address, a floating-point address, or a constant.

The destination is a word address or a floating-point address that stores the result. If the destination is a word address other than a floating-point and the result is greater than 32,767, then the value 32,767 is placed in the result. However, if the controller is a Series C or later 5/02 or 5/03, 5/04, 5/05 and has the S:2/14 (math overflow selection bit) set, then the unsigned, truncated least significant 16 bits of the overflow remain in the destination.

ControlLogix.

Source A may be a SINT, INT, DINT, or REAL value, which is divided by source B.

Source B may be a SINT, INT, DINT, or REAL value, by which source A is divided.

The destination is a SINT, INT, DINT, or REAL tag address, which stores the result.

DINT and REAL are the optimal data types. If source B is zero, the destination is set to the value at source A.

Figure 6.8 DIV instruction.

6.2.6 Modulo (MOD) Instruction

The modulo (MOD) instruction (ControlLogix only) is an output instruction that divides source A by source B and stores the remainder in the destination address.

In Figure 6.9 the value stored at the address of source A is divided by the value stored in source B, and the remainder is stored in the destination address. For example, $5 \div 2 = 2$ with a remainder of 1, which is the value stored.

ControlLogix.

> Source A may be a SINT, INT, DINT, or REAL value, which is divided by source B.
>
> Source B may be a SINT, INT, DINT, or REAL value, by which source A is divided.
>
> The destination is a SINT, INT, DINT, or REAL tag address that stores the remainder. DINT and REAL are the optimal data types. If source B is 0, the setting of the destination depends on the tag type. If the tag type is SINT, INT, or DINT, the destination is set to 0. If the tag type is REAL, the destination is set to infinity.

6.2.7 Square Root (SQR) Instruction

The square root (SQR) instruction is an output instruction that calculates the square root of the source and places the result in the destination address.

In Figure 6.10 the square root of the value stored at the address of the source is calculated, and the result is stored in the destination address.

SLC 500.

> The source may be a word address or a floating-point address.
>
> The destination is a word address or a floating-point address that stores the result of the square root calculation.

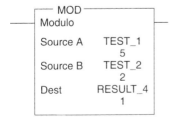

Figure 6.9 MOD instruction.

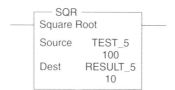

Figure 6.10   SQR instruction.

ControlLogix.

The source may be a SINT, INT, DINT, or REAL value, of which the square root is calculated.

The destination is a SINT, INT, DINT, or REAL tag address that stores the result.

DINT and REAL are the optimal data types.

6.2.8 Negate (NEG) Instruction

The negate (NEG) instruction is an output instruction that changes the sign of the source and places the result in the destination address.

In Figure 6.11 the sign of the value stored at the address of source is changed, and the result is stored in the destination address.

SLC 500.

The source may be a word address or a floating-point address.

The destination is a word address or a floating-point address that stores the result of the negate operation.

ControlLogix.

The source may be a SINT, INT, DINT, or REAL value, which is negated.

The destination is a SINT, INT, DINT, or REAL tag address that stores the result.

DINT and REAL are the optimal data types.

6.2.9 Absolute (ABS) Instruction

The absolute (ABS) instruction is an output instruction that places the absolute value of the source in the destination address, as illustrated in Figure 6.12.

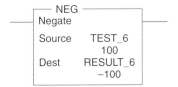

Figure 6.11 NEG instruction.

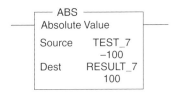

Figure 6.12 ABS instruction.

SLC 500.

> The source may be a word address or a floating-point address.
>
> The destination is a word address or a floating-point address that stores the result of the ABS operation.

ControlLogix.

> The source may be a SINT, INT, DINT, or REAL value.
>
> The destination is a SINT, INT, DINT, or REAL tag address that stores the absolute value of the source. DINT and REAL are the optimal data types.

6.2.10 Clear (CLR) Instruction

The Clear (CLR) instruction is an output instruction that clears (zeros) all the bits stored in the destination address.

In Figure 6.13 the bits stored at the destination address are cleared.

SLC 500.

> The destination is a word address or a floating-point address that stores the result of the clear operation.

ControlLogix.

> The destination is a SINT, INT, DINT, or REAL tag address that stores the result of the clear operation. DINT and REAL are the optimal data types.

Figure 6.13 CLR instruction.

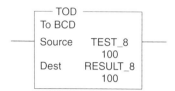

Figure 6.14 TOD instruction.

6.2.11 Convert to BCD (TOD) Instruction

The convert to BCD (TOD) instruction is an output instruction that converts the decimal value of the source to binary-coded decimal (BCD) and places the result in the destination address.

In Figure 6.14 the value stored at the address of source is converted to BCD and the BCD result is stored in the destination address.

SLC 500.

The source may be a word address. A floating-point address is not allowed. The range is 0 to 9999.

The destination is a word address. A floating-point address is not allowed. The address stores the result of the TOD operation. If the integer value entered is negative, the sign is ignored and the conversion occurs as if the number were positive.

ControlLogix.

The source may be a SINT, INT, or DINT value, which is converted to BCD. The range is 0 to 99,999,999.

The destination is a SINT, INT, or DINT tag address that stores the result.

DINT is the optimal data type. A negative value at the source will clear the destination.

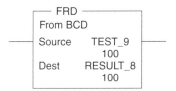

Figure 6.15 FRD instruction.

6.2.12 Convert from BCD (FRD) Instruction

The convert from BCD (FRD) instruction is an output instruction that converts the BCD value of the source to decimal and places the result in the destination address.

In Figure 6.15 the value stored at the address of the source is converted from BCD to decimal, and the decimal result is stored in the destination address.

SLC 500.

The source may be a word address. A floating-point address is not allowed.

The destination is a word address. A floating-point address is not allowed. The address stores the result of the FRD operation.

ControlLogix.

The source may be a SINT, INT, or DINT value, which is converted to decimal.

The destination is a SINT, INT, or DINT tag address that stores the result.

DINT is the optimal data type.

▶ 6.3 COMPARISON INSTRUCTIONS

Comparison instructions are input instructions that do comparisons either by using an expression (only in ControlLogix) or doing the comparison indicated by the specific instruction. Comparisons may be done between different data table types, such as between integer and floating-point types.

6.3.1 Compare (CMP) Instruction

The compare (CMP) instruction (ControlLogix only) is an input instruction that does a comparison on the arithmetic operations specified by the expression.

The expression may contain arithmetic operators, comparison operators, tags, or immediate values. Parentheses are used to define sections when creating a complex expression.

Execution time of a CMP instruction is slower and uses more memory than the execution time of the sum of the individual compare instructions. The advantage of a CMP instruction is that it allows multiple operations to be done in one instruction.

The comparison instruction shown in Figure 6.16 is logically true when the sum of data3 and data4 is equal to 15.

ControlLogix.

The expression is a SINT, INT, DINT, or REAL data type. DINT and REAL are the optimal data types. A SINT or INT converts to a DINT value by sign extension.

In Figure 6.17, there is no comparison operator in the expression. In the case where there is no comparison operator in the expression, the instruction is true if the value of the expression is nonzero and false if the value of the expression is zero.

6.3.2 Equal (EQU) Instruction

The equal (EQU) instruction is an input instruction that compares source A to source B. When source A is equal to source B, the instruction is logically true; otherwise, it is logically false.

In Figure 6.18 the value stored at the address of source A is compared to the value stored at source B. If the values are equal, the instruction is logically true. If they are unequal, the instruction is logically false.

Figure 6.16   CMP instruction.

Figure 6.17   CMP with no comparison operator.

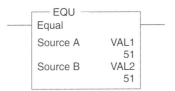

Figure 6.18 EQU instruction.

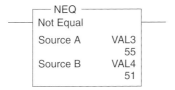

Figure 6.19 NEQ instruction.

SLC 500.

Source A may be a word address or a floating-point address.

Source B may be a word address, a floating-point address, or a constant value.

ControlLogix.

Source A may be SINT, INT, DINT, or REAL.

Source B may be SINT, INT, DINT, or REAL. DINT and REAL are the optimal data types. SINT or INT tags convert to a DINT value by sign extension.

6.3.3 Not Equal (NEQ) Instruction

The not equal (NEQ) instruction is an input instruction that compares source A to source B. When source A is not equal to source B, the instruction is logically true; otherwise, it is logically false.

In Figure 6.19 the value stored at the address of source A is compared to the value stored at source B. If the values are not equal, the instruction is logically true. If they are equal, the instruction is logically false.

SLC 500.

Source A may be a word address or a floating-point address.

Source B may be a word address, a floating-point address, or a constant value.

ControlLogix.

> Source A may be SINT, INT, DINT, or REAL.
>
> Source B may be SINT, INT, DINT, or REAL. DINT and REAL are the optimal data types. SINT or INT tags convert to a DINT value by sign extension.

6.3.4 Less Than (LES) Instruction

The less than (LES) instruction is an input instruction that compares source A to source B. When source A is less than source B, the instruction is logically true; otherwise, it is logically false.

In Figure 6.20 the value stored at the address of source A is compared to the value stored at source B. If source A is less than source B, the instruction is logically true. If they are equal or source A is greater than source B, the instruction is logically false.

SLC 500.

> Source A may be a word address or a floating-point address.
>
> Source B may be a word address, a floating-point address, or a constant value.

ControlLogix.

> Source A may be SINT, INT, DINT, or REAL.
>
> Source B may be SINT, INT, DINT, or REAL. DINT and REAL are the optimal data types. SINT or INT tags convert to a DINT value by sign extension.

6.3.5 Less Than or Equal (LEQ) Instruction

The less than or equal (LEQ) instruction is an input instruction that compares source A to source B. When source A is less than or equal to source B, the instruction is logically true; otherwise, it is logically false.

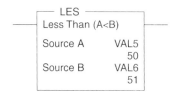

Figure 6.20   LES instruction.

In Figure 6.21 the value stored at the address of source A is compared to the value stored at source B. If source A is less than or equal to source B, the instruction is logically true. If source A is greater than source B, the instruction is logically false.

SLC 500.

Source A may be a word address or a floating-point address.

Source B may be a word address, a floating-point address, or a constant value.

ControlLogix.

Source A may be SINT, INT, DINT, or REAL.

Source B may be SINT, INT, DINT, or REAL. DINT and REAL are the optimal data types. SINT or INT tags convert to a DINT value by sign extension.

6.3.6 Greater Than (GRT) Instruction

The greater than (GRT) instruction is an input instruction that compares source A to source B. When source A is greater than source B, the instruction is logically true, otherwise, it is logically false.

In the example in Figure 6.22 the value stored at the address of source A is compared to the value stored at source B. If source A is greater than source B, the instruction is logically true. If source A is equal to or less than source B, the instruction is logically false.

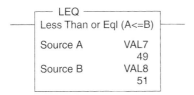

Figure 6.21 LEQ instruction.

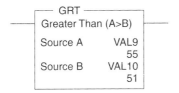

Figure 6.22 GRT instruction.

SLC 500.

Source A may be a word address or a floating-point address.

Source B may be a word address, a floating-point address, or a constant value.

ControlLogix.

Source A may be SINT, INT, DINT, or REAL.

Source B may be SINT, INT, DINT, or REAL. DINT and REAL are the optimal data types. SINT or INT tags convert to a DINT value by sign extension.

6.3.7 Greater Than or Equal (GEQ) Instruction

The greater than or equal (GEQ) instruction is an input instruction that compares source A to source B. When source A is greater than or equal to source B, the instruction is logically true; otherwise, it is logically false.

In Figure 6.23 the value stored at the address of source A is compared to the value stored at source B. If source A is greater than or equal to source B, the instruction is logically true. If source A is less than source B, the instruction is logically false.

SLC 500.

Source A may be a word address or a floating-point address.

Source B may be a word address, a floating-point address, or a constant value.

ControlLogix.

Source A may be SINT, INT, DINT, or REAL.

Source B may be SINT, INT, DINT, or REAL. DINT and REAL are the optimal data types. SINT or INT tags convert to a DINT value by sign extension.

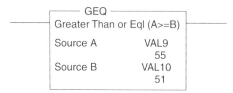

Figure 6.23   GEQ instruction.

6.3.8 Mask Equal (MEQ) Instruction

The mask equal (MEQ) instruction is an input instruction that compares the source to the compare through a mask. A 1 in the mask means the bit status in the source at that location will be compared to the bit status at the same location in the compare. A 0 in the mask means the bit status in the source at that location will not be compared to the bit status at the same location in the compare. When the source matches the compare at all the bit locations where there is a 1 in the mask, the instruction will be true; otherwise, it will be false.

In Figure 6.24 the value stored at the address of the source is compared to the value stored at the compare through the mask. If the source is equal to the compare at the bit locations where there is a 1 in the mask, the instruction is logically true. Otherwise, the instruction is logically false. The shaded bits in the source are compared to the shaded bits in the compare.

SLC 500.

Source may be a word address to which the value in compare is being compared.

Mask may be a word address or hexadecimal value. It may be entered in hexadecimal by entering the value followed by the letter *h*, in binary by typing in the value followed by the letter *b*, or in decimal. It will always be displayed in hexadecimal. Bit status determines whether to block or allow comparison.

Compare may be a word address or integer value. Its value is compared to the source value.

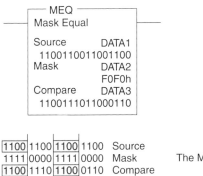

Figure 6.24 MEQ instruction.

ControlLogix.

Source may be a SINT, INT, or DINT value, which is compared to the compare value.

Mask may be a SINT, INT, or DINT value. It contains bit status that blocks or allows comparison. It defaults to entering the value in decimal. To enter the value in hexadecimal, precede the number with #16, for octal precede the number with #8, and for binary precede the number with #2.

Compare may be a SINT, INT, or DINT value, which is compared to the source value. DINT is the optimal data type. SINT or INT tags convert to a DINT value by zero fill.

6.3.9 Limit (LIM) Instruction

The limit (LIM) instruction is an input instruction that compares the test to the low limit and the high limit. The instruction may test between (and including) the limits or outside of (and including) the limits, depending on the value relationship between the low limit and the high limit. If the high limit is greater to or equal to the low limit, the instruction is true when the test value is between or equal to the limits. If the high-limit value is less than or equal to the low limit, the instruction is true when the test value is outside of (and including) the limits.

In Figure 6.25 the value stored at the address of the test is compared to the values stored at the low limit and the high limit. In this example, if the test value is between (and including) the limits (100–200), then the instruction is true, which is the situation in the example, because 150 is between 100 and 200.

In Figure 6.26 the value stored at the address of the test is compared to the values stored at the low limit and the high limit. In this example, if the test value is equal to or greater than 200 or equal to or less than 100, the instruction is true. If the test value is between 101 and 199, inclusive, the instruction is false.

SLC 500.

Low limit may be a word address if the test value is a constant or may be a word address or a constant if the test value is a word address.

Test may be a word address or a constant.

High limit may be a word address if test is a constant, or it may be a word address or a constant if test is a word address.

ControlLogix.

Low limit may be SINT, INT, DINT, or REAL.

Test may be SINT, INT, DINT, or REAL.

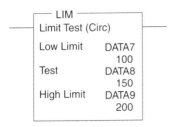

Figure 6.25 LIM instruction (low limit < high limit).

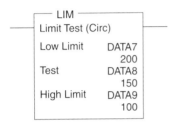

Figure 6.26 LIM instruction (low limit > high limit).

High limit may be SINT, INT, DINT, or REAL. DINT or REAL are the optimal data types. SINT or INT tags convert to a DINT value by sign extension.

▶ 6.4 LOGICAL INSTRUCTIONS

Logical instructions are output instructions that do logical operations on bits. In an SLC 500 controller, the logical operation is done between two words. In a ControlLogix processor the number of bits upon which the logical operation is done depends on the data type.

6.4.1 Bitwise AND (AND) Instruction

The bitwise AND (AND) instruction is an output instruction that performs a bit-by-bit AND operation between source A and source B and stores the result in the destination.

Table 6.1 shows the different AND operations on a bit-by-bit basis. The only time a 1 appears in the destination is when a 1 appears in both source A and source B.

Figure 6.27 illustrates the operation of the AND instruction. An AND operation is done between the bits in source A and the bits in source B on a bit-by-bit basis, and the result is stored in the destination. This operation is executed during every scan while the instruction is true.

TABLE 6.1   BITWISE AND INSTRUCTION TRUTH TABLE

Source A	Source B	Destination
0	0	0
1	0	0
0	1	0
1	1	1

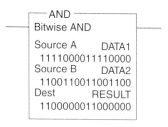

Figure 6.27   AND instruction.

SLC 500.

Source A is a word, on which the AND operation is performed with source B.

Source B is a word address or a constant, on which the AND operation is performed with source A. Both source A and source B cannot be constants.

The destination is a word address that stores the result of the AND operation between source A and source B.

ControlLogix.

Source A may be a SINT, INT, or DINT value, on which the AND operation is performed with source B.

Source B may be a SINT, INT, or DINT value, on which the AND operation is performed with source A.

Destination may be SINT, INT, or DINT. It stores the result of the AND operation between source A and source B. DINT is the optimal data type. SINT or INT tags convert to a DINT value by zero fill.

6.4.2 Bitwise OR (OR) Instruction

The bitwise OR (OR) instruction is an output instruction that performs a bit-by-bit OR operation between source A and source B and stores the result in the destination.

TABLE 6.2 OR INSTRUCTION TRUTH TABLE

Source A	Source B	Destination
0	0	0
1	0	1
0	1	1
1	1	1

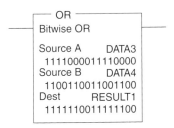

Figure 6.28   OR instruction.

Table 6.2 shows the different OR operations on a bit-by-bit basis. The only time a 0 appears in the destination is when a 0 appears in both source A and source B.

Figure 6.28 illustrates the operation of the OR instruction. An OR operation is done on the bits in source A with the bits in source B on a bit-by-bit basis, and the result is stored in the destination. This operation is executed every scan while the instruction is true.

SLC 500.

Source A is a word address, on which the OR operation is performed with source B.

Source B is a word address or a constant, on which the OR operation is performed with source A. Both source A and source B cannot be constants.

The destination is a word address that stores the result of the OR operation between source A and source B.

ControlLogix.

Source A may be a SINT, INT, or DINT value on which the OR operation is performed with source B.

Source B may be a SINT, INT, or DINT value, on which the OR operation is performed with source A.

Destination may be SINT, INT, or DINT. It stores the result of the OR operation between source A and source B. DINT is the optimal data type. SINT or INT tags convert to a DINT value by zero fill.

6.4.3 Bitwise Exclusive OR (XOR) Instruction

The bitwise exclusive OR (XOR) instruction is an output instruction that performs a bit-by-bit XOR operation between source A and source B and stores the result in the destination.

Table 6.3 shows the different XOR operations on a bit-by-bit basis. The only time that a 1 appears in the destination is when a difference occurs in the status of the bits in source A and source B.

Figure 6.29 illustrates the operation of the XOR instruction. An XOR operation is performed on the bits in source A and the bits in source B on a bit-by-bit basis. The result is stored in the destination. This operation is executed during every scan while the instruction is true.

SLC 500.

Source A is a word, on which the XOR operation is performed with source B.

Source B is a word address or a constant, on which the XOR operation is performed with source A. Source A and source B cannot both be constants.

The destination is a word address that stores the result of the XOR operation between source A and source B.

TABLE 6.3 BITWISE XOR INSTRUCTION TRUTH TABLE

Source A	Source B	Destination
0	0	0
1	0	1
0	1	1
1	1	0

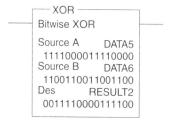

Figure 6.29 XOR Instruction

ControlLogix.

Source A may be a SINT, INT, or DINT value, on which the XOR operation is performed with source B.

Source B may be a SINT, INT or DINT value, on which the XOR operation is performed with source A.

The destination may be SINT, INT, or DINT, and it stores the result of the XOR operation between source A and source B. DINT is the optimal data type. SINT or INT tags convert to a DINT value by zero fill.

6.4.4 Bitwise NOT (NOT) Instruction

The bitwise NOT (NOT) instruction is an output instruction that performs a bit-by-bit NOT operation on the data stored in the source and stores the result in the destination.

Table 6.4 shows the different NOT operations on a bit-by-bit basis. The status of each bit is inverted from the source to the destination.

Figure 6.30 illustrates the operation of the NOT instruction. The bits in the source are inverted on a bit-by-bit basis, and the result is stored in the destination. This operation is executed during every scan while the instruction is true.

SLC 500.

Source is a word on which the NOT operation is performed.

The destination is a word address that stores the result of the NOT operation of the source.

TABLE 6.4   BITWISE NOT INSTRUCTION TRUTH TABLE

Source	Destination
0	1
1	0

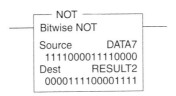

Figure 6.30   NOT instruction.

ControlLogix.

Source is a SINT, INT, or DINT value, on which the NOT operation is performed.

Destination is SINT, INT, or DINT, and it stores the result of the NOT operation of the source. DINT is the optimal data type. SINT or INT tags convert to a DINT value by zero fill.

SUMMARY

Data-manipulation instructions are output instructions that copy data from one memory location or a constant value to another memory location. Arithmetic instructions are output instructions that do mathematical operations between two memory locations or a memory location and a constant and store the result at another memory location. Comparison instructions are input instructions that compare data between two memory locations or a memory location and a constant. If the comparison is true, the instruction is true. Logical instructions are output instructions that do logical operations between two memory locations and store the result in another memory location.

► EXERCISES

1. Describe the function of a MOV instruction.
2. When does the MOV instruction execute?
3. Describe the function of the mask in an MVM instruction.
4. Describe the function of the FRD instruction.
5. What instruction can be used to zero the value at an address?
6. What instruction can be used to turn on an alarm if the pressure goes above 300 psi or below 100 psi? Show how the instruction would be set up.
7. What instruction can be used to compare the bit status between two memory locations and store in another memory location where the bit status was different between the two memory locations?
8. A process has to have an oven temperature set at one of two set points, depending on the type of part being placed in the oven. Design the ladder logic that will copy the data from either SETPOINT_1 or SETPOINT_2 to OVEN_1. It will copy SETPOINT_1 when switch SW_1 is closed and will copy SETPOINT_2 when SW_1 is open.
9. On a control panel, there is a two-digit LED display. The display is a BCD display and is connected to the first eight terminals (0–7) on a 16-point module. Design logic that writes the value stored at DATA_1 in a decimal format, with

a range of 0–99, to the LED display and allows other devices to be connected to the upper eight terminals (8–15) on the output card. The address of the 16 bits on the card is OUTPUT_1.

10. Write the expression for a compute instruction that will first divide the value stored at DATA_2 by 25 and then add 15 to the result.

11. Do the same operation as in Exercise 10, except use individual instructions for the math operations.

12. For a ControlLogix controller, set up an instruction that divides DATA_2 by 25 and stores the remainder at DATA_3.

13. Design ladder logic that turns on an output, OUTPUT_5, when bits 2, 4, 7, 13, and 15 of DATA_6 match the corresponding bits in DATA_7.

14. Set up an instruction that compares the bits in VALUE_1 to the bits in VALUE_2 and puts a 1 in the bit locations in VALUE_3 where there is a difference in the status of the bits between VALUE_1 and VALUE_2.

7 SLC 500 and ControlLogix File-Data Manipulation

CHAPTER OUTLINE

LEARNING OBJECTIVES

Upon reading this chapter students should be able to:

- Explain data file- (array-) level concepts.
- Explain file- (array-) level instruction structure.
- Interpret and program file- (array-) level instructions.

INTRODUCTION

This chapter defines the structure and operation of various file-level instructions in SLC 500 and ControlLogix controllers. The small differences in the function of the instructions between the controllers are pointed out where they occur.

In SLC 500 controllers a group of consecutively addressed words or elements is called a file, whereas in ControlLogix controllers the same structure is referred to as an array. Chapter 3 discussed the difference in addressing between SLC 500 controllers and ControlLogix controllers.

The file-level instructions include the instructions that do arithmetic, logic, shift, and function operations on values in files or arrays.

File-level instructions do operations similar to word-level instructions, but instead of acting on a word of data, they operate on groups of words, referred to as files (SLC 500) or arrays (ControlLogix).

File-level instructions may operate on data file-to-file, word-to-file, or file-to-word. Figure 7.1 illustrates each method.

The length of the file or array is indicated in the instruction, and the way the address is entered determines whether the instruction acts on a single element or on a file or an array. In the SLC 500 controller a file-level address is designated by placing a # sign at the beginning of the address. Thus an address such as #N22:5 indicates a file address with a beginning address of N22:5. The length of the file is indicated in the instruction. In a ControlLogix controller an array tag, such as data_1[5], is used. This is the starting address, and the length is indicated in the instruction. An example of an element address in the SLC 500 is N22:15 and an example of an element address in ControlLogix is pressure_1.

When entering an instruction, depending on the type of file-level instruction, the mode of operation may also have to be entered. The mode of operation determines how the data are distributed when the instruction is executed. There are three choices for the mode: the All mode, the Numeric mode, and the Incremental mode. Table 7.1 shows the operation for the different modes.

Figure 7.2 shows the order in which the data are acted upon in the All mode. All five words will be acted on, and the done bit for the instruction will be set before the scan continues on to the next instruction.

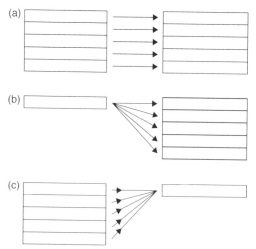

Figure 7.1 Data Operation; (a) File-to-file or array-to-array; (b) Word-to-file or element-to-array; (c) File-to-word or array-to-element.

TABLE 7.1 MODES OF OPERATION

Mode	Operation
All	Operates on all the words or elements specified by the length before the scan continues to the next instruction
Numeric	Operates only on the number of words or elements entered in the mode before the scan continues to the next instruction
Incremental	Operates on one element per false-to-true transition of the instruction

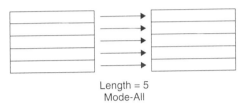

Length = 5
Mode-All

Figure 7.2 All mode.

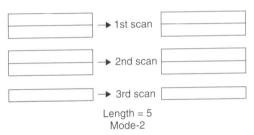

1st scan

2nd scan

3rd scan

Length = 5
Mode-2

Figure 7.3 Numeric mode.

The Numeric mode is shown in Figure 7.3. The mode is 2, so when the instruction goes true, in the first scan the instruction will operate on the first two words of data before it goes to the next instruction. In the second scan it will operate on the next two words of data and then continue on to the next instruction. On the third scan it will operate on the remaining one word and then go on to the next instruction. It will operate on all five words in one false-to-true transition of the instruction and take three scans to do it. Even if the instruction goes false after the first scan, the instruction will continue its operation until it is done.

Figure 7.4 illustrates the sequence of data operation when the mode selection is Incremental. One word is operated on per false-to-true transition of the instruction. In the example of Figure 7.4, it will take five false-to-true transitions for all the data to be operated on. In the All and Numeric modes, it only takes one false-to-true transition for all the data to be operated on.

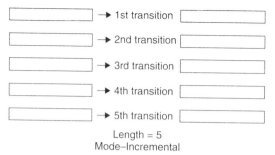

Length = 5
Mode–Incremental

Figure 7.4 Incremental mode.

▶ 7.2 FILE ARITHMETIC AND LOGIC (FAL) INSTRUCTION

The file arithmetic and logic (FAL) instruction (ControlLogix only) is an output instruction that operates on elements and arrays of data. The instruction may perform array-to-array, element-to-array, or array-to-element operations, as described in Section 7.1. The expression may be structured the same as the expression in the CPT instruction. The operators in an expression may be +, −, *, /, **· ABS, ACS, AND, ASN, ATN, COS, DEG, GRD, LN, LOG, MOD, NOT, OR, RAD, SIN, SQR, TAN, TOD, TRN, and XOR. Each operator in an expression must have one or two operands (tags or immediate values). There is no limit to the length of the expression. An example is

operator(operand)   or   operand_1 operator operand_2

The order of operations for the instructions is shown in Table 7.2. Operations of equal order are performed from left to right.

Table 7.3 shows the data structure of the FAL instruction. For the destination and the expression, the optimal data types are DINT and REAL.

Table 7.4 lists the components of the control structure for the FAL instruction.

7.2.1 FAL Array-to-Array Copy

The FAL instruction is shown in Figure 7.5. The FAL is an output instruction. This is an example of an array-to-array copy, because both the expression and the destination addresses are array-level addresses. When the instruction goes from false to true, the instruction will copy all seven elements from the expression to the destination in one scan because the instruction is in the All mode.

TABLE 7.2 ORDER OF OPERATIONS FOR THE EXPRESSION

Order	Operation
1	( )
2	ABS, ACS, ASN, ATN, COS, DEG, FRD, LN, LOG, RAD, SIN, SQR, TAN, TOD, TRN
3	**
4	– (negate), NOT
5	*, /, MOD
6	– (subtract), +
7	AND
8	XOR
9	OR

TABLE 7.3 FAL INSTRUCTION OPERANDS

Operand	Data	Format	Description
Control	CONTROL	Tag	Control structure for the operation
Length	DINT	Immediate	The number of elements on which the instruction operates
Position	DINT	Immediate	Current element in the array
Mode	DINT	Immediate	The method of operation on the data
Destination	SINT INT DINT REAL	Tag	Where the result is stored
Expression	SINT	Immediate	Consists of tages and/or immediate values separated by operators
	INT DINT REAL	Tag	

7.2.2 FAL Element-to-Array Copy

Figure 7.6 illustrates a FAL instruction doing an element-to-array copy. In this example the expression is an element address and the destination is an array address. Because the mode is 1, it will take one false-to-true transition of the instruction and eight scans for the instruction to complete its operation.

TABLE 7.4 CONTROL STRUCTURE

Mnemonic	Data Type	Description
.en	BOOL	Enable bit, set when FAL is enabled.
.dn	BOOL	Done bit, set when the instruction has operated on the last element.
.er	BOOL	Error bit, set when there is an overflow created by the expression. The instruction stops executing at that position, and the .er bit must be reset for the operation of the instruction to continue.
.len	DINT	Length, the number of elements upon which instruction will operate.
.pos	DINT	Position, the current element upon which the instruction is operating.

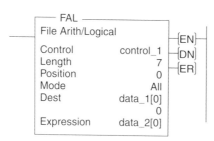

Figure 7.5 FAL array-to-array copy.

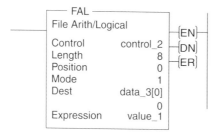

Figure 7.6 Element-to-array copy.

7.2.3 FAL Array-to-Element Copy

Figure 7.7 illustrates the FAL configured for an array-to-element copy in the Incremental mode. As the instruction is toggled, it indexes through the seven elements of the array. It takes seven transitions before the last value in the array is

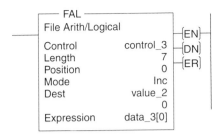

Figure 7.7 FAL array-to-element copy.

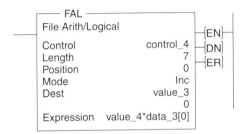

Figure 7.8 FAL arithmetic operation.

Figure 7.9 Element times array copied to element.

transferred to the destination. Each time the instruction is indexed, the previous value stored in value_2 is overwritten with a new value.

7.2.4 FAL Arithmetic Operation

The FAL instruction in Figure 7.8 is configured to do an arithmetic operation in which the values in an array are multiplied by the value at an element address and the result is stored in an element address. The mode for the instruction is set at Incremental, so a single multiplication takes place every time the instruction is transitioned from false to true. Each time the instruction is enabled, the previous value in the destination is overwritten.

Figure 7.9 illustrates the data transfer of the instruction shown in Figure 7.8.

▶ 7.3 COPY FILE (COP) INSTRUCTION

The Copy File (COP) instruction is an output instruction that is available in both the SLC 500 and ControlLogix controllers. Its function is similar to that of the FAL doing an array-to-array copy in the All mode. The COP instruction executes every scan when its rung is true.

The COP for the SLC 500 is shown in Figure 7.10. In this example the data in words N7:50 through N7:59 are copied into N20:5 through N20:14. This is done when the instruction is enabled. Both the source and the destination addresses must be file-level addresses (the address starts with the # sign). The maximum length is 128 elements.

Figure 7.11 illustrates the COP instruction used with the ControlLogix controller. In this example, the data in the first ten elements of the array test_1 are copied into the first ten words of the array test_2. The COP instruction operates on contiguous memory locations and performs a straight byte-to-byte copy. If the byte count is greater than the length of the source, unpredictable data are copied to the remaining elements. Also, the COP instruction does not write past the end of an array. If the length is greater than the length of the destination array, the COP instruction stops at the end of the array without a major fault being generated. The source and the destination should be the same data types, or unexpected results may occur.

Table 7.5 shows the operand structure for the ControlLogix COP instructions. For the source and the destination, the optimal data types are SINT, INT, DINT, and REAL.

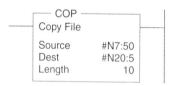

Figure 7.10 SLC 500 COP instruction.

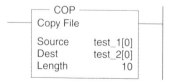

Figure 7.11 ControlLogix COP instruction.

TABLE 7.5 CONTROLLOGIX COP INSTRUCTION OPERANDS

Operand	Type	Format	Description
Source	SINT INT DINT REAL Structure	Tag	Starting address of the data to copy
Destination	SINT INT DINT REAL Structure	Tag	Starting address of the data to be overwritten
Length	DINT	Immediate tag	Number of elements in the destination to copy

▶ 7.4 FILE FILL (FLL) INSTRUCTION

The File Fill (FLL) instruction is an output instruction available in both the SLC 500 and ControlLogix controllers. Its function is similar to the FAL doing an element-to-array copy in the All mode. The FLL instruction executes every scan when its rung is true.

The FLL instruction for the SLC 500 is shown in Figure 7.12. In this example the data in word N13:0 are copied into N14:0 through N14:9. This is done when the instruction is enabled. The source is an element address, and the destination addresses must be file-level addresses (the address starts with the # sign). The maximum length is 128 elements.

Figure 7.13 illustrates the FLL instruction when it is used with the Control-Logix controller. In this example, the data in element of value_5 are copied into the first ten words of the array test_4[0]. The FLL instruction fills the array with the value of the source. The number of bytes filled in the destination is equal to the length times the number of bytes in the destination data type. Also, the FLL

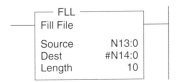

Figure 7.12   SLC 500 FLL instruction.

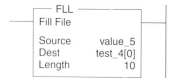

Figure 7.13 ControlLogix FLL instruction.

TABLE 7.6 CONTROLLOGIX FLL INSTRUCTION OPERANDS

Source	Destination	Source Converted To
SINT, INT, DINT, or REAL	SINT	SINT
SINT, INT, DINT, or REAL	INT	INT
SINT, INT, DINT, or REAL	DINT	DINT
SINT, INT, DINT, or REAL	REAL	REAL
SINT	Structure	SINT (not converted)
INT	Structure	INT (not converted)
DINT	Structure	DINT (not converted)
REAL	Structure	REAL (not converted)

instruction does not write past the end of an array. If the length is greater than the length of the destination array, the FLL instruction stops at the end of the array without a major fault being generated. Table 7.6 shows the operand types for the FLL instruction. The source and the destination should be the same data types. If the source and destination are different data types, the destination elements will be filled with converted source values.

SUMMARY

The file-level instructions allow us to operate on groups of data with a single instruction. The FAL instruction can be used to copy data, do math operations on data, and do logical operations on data. The COP and FLL instructions do operations that can be done with the FAL instruction, but they do it faster and more efficiently than the FAL instruction. The COP copies a group of data from one memory location to another, and the FLL fills a file with a single value. A group of consecutive elements in an SLC 500 are referred to as a file, and in a ControlLogix controller they are referred to as an array.

► EXERCISES

1. To have the FAL instruction operate on the complete array of data in one scan, the mode would have to be what?

2. The FAL instruction is available in what controller(s)?

3. What is the format for a file address in the SLC 500 controller?

4. What is the format for an array address in the ControlLogix controller?

5. Set up a FAL instruction to divide the data in the first ten elements of the array data_1[0] by 15 and store the result in the first ten elements of the array data_2[10]. Have it operate on two elements of data per scan.

6. Set up a FAL instruction that will copy the data out of ten consecutive elements, starting with value_1, into the single element value_20, one element per false-to-true transition of the FAL instruction.

7. A facility has five ovens. For each oven the operator must be able to control the temperature set points between idle and run by toggling a switch SW_1 between the idle position and the run position. Design the ladder logic that will accomplish this.

8. Add the logic to Exercise 4 that will zero the set points of the ovens when switch SW_4 is turned on.

8

SLC 500 and ControlLogix Shift Register and Sequencer Instructions

CHAPTER OUTLINE

LEARNING OBJECTIVES

Upon reading this chapter students should be able to:

- Explain bit-shift concepts.
- Interpret and program bit-shift instructions.
- Explain first-in/first-out and last-in/first-out shift-register concepts.
- Interpret and program FIFO load and FIFO unload instructions.
- Interpret and program LIFO load and LIFO unload instructions.
- Explain sequencer concepts.
- Interpret and program SQI, SQO, and SQL instructions.

INTRODUCTION

This chapter defines the structure and operation of various shift-register and sequencer instructions in the SLC 500 and ControlLogix controllers. The structure of the instructions in the SLC 500 and ControlLogix controllers is similar, except for the sequencers.

The bit-shift, first-in/first-out, and last-in/first-out instructions are frequently used as tracking tools, for example to track parts on an assembly line. The sequencer instructions are used in the control of sequential-type operations where the process cycles in the same repeated sequence.

▶ 8.1 BIT-SHIFT INSTRUCTIONS

Bit-shift instructions are used for tracking on a bit level. Because a bit can store only a 0 or a 1, we can track only whether a part is present/not present or a good part/bad part. The instructions shift the status of bits through an array or file. By examining these bits, we can control outputs to do the required function, such as filling a container if the container is present or rejecting a bad part.

8.1.1 Bit-Shift-Left (BSL) Instruction

Figure 8.1 shows the bit-shift-left (BSL) instruction for the SLC 500 controller. The following parameters need to be entered in the SLC 500 BSL instruction:

File	The address of the bit file to be manipulated. The address must start with a # sign. It starts at bit 0 of the element. Any remaining bits in the last element of the file address cannot be used because the instruction invalidates them.
Control	R data-table type. The address is unique to the instruction. It is a three-word element that consists of the status word, the length, and the position.
Bit address	An address input to the shift register. Its status is shifted into the first bit of the file.
Length	File length in bits. The maximum value is 2048 bits.

The status bits of the control word are the enable (en), the done (dn), the error (er), and the unload (ul) bits. The enable bit follows the instruction's status, the done bit is set when the instruction has shifted all bits in the file one position, the error bit is set when an error occurs in the operation of the instruction (negative value for the length), and the unload bit's status is controlled by shifting the status of the last bit of the file into the unload bit when the instruction is executed.

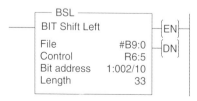

Figure 8.1 SLC 500 bit-shift-left (BSL) instruction.

The BSL instruction for the ControlLogix controller, shown in Figure 8.2, has the same operands as the BSL in the SLC 500 controller. The operands for the instruction are shown in Table 8.1.

The control structure for the BSL in the ControlLogix is the same as the control structure for the SLC 500.

Figure 8.3 shows the operation of BSL instruction for the SLC 500 controller. The operation for the BSL instruction in the ControlLogix controller is the same.

When the BSL instruction is enabled, the processor sets the en bit and then shifts the 33 bits in bit file B9:0 one bit position. The last bit shifts out of position 33 into the ul bit. The source bit, Bit address I:002/10, shifts into the first bit position, B9/00.

If the preceding example were tracking the presence of parts on a 33-station assembly line, instead of 33 sensors on the line to determine the presence of parts, a sensor would be used to determine if a part were placed on the line. From that point on, the shift register would track the parts. For example, to determine if there were a part at station 12, bit address B9/12 would be examined and the appropriate action would be taken.

8.1.2 Bit-Shift-Right (BSR) Instruction

The bit-shift-right (BSR) instruction has the same operands as the BSL instruction. The difference is the direction in which the bits are indexed.

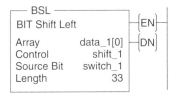

Figure 8.2 ControlLogix bit-shift-left instruction.

TABLE 8.1   OPERANDS FOR THE CONTROLLOGIX BSL INSTRUCTION

Operand	Data	Format	Description
Array	DINT	Array tag	Address of the first element of the array through which the bits will be shifted
Control	CONTROL	Tag	Control structure for the instruction
Source bit	BOOL	Tag	Bit shifted into the first bit of the array
Length	DINT	Immediate	Number of bits of the array

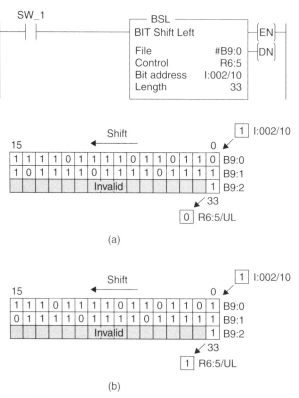

Figure 8.3 Operation of the BSL instruction; (a) Before instruction enabled; (b) After instruction enabled.

Figures 8.4 and 8.5 show the BSR for the SLC 500 and ControlLogix controllers. The BSR instruction shifts all bits within the file or array one position to the right (from a higher address to a lower address). The bit address shifts into the last bit position in the file, and the bit status from the first bit shifts into the ul bit.

Figure 8.6 shows the operation of the BSR instruction. When the BSR instruction is enabled, the processor sets the en bit and then shifts the 35 bits in bit file B9:100 one bit position. The first bit shifts out of position 35 into the ul bit. The source bit, bit address I:002/10, shifts into the last bit position, B9/1619 (B9:102/02).

8.1.3 FIFO Load (FFL) and FIFO Unload (FFU) Instructions

The first-in/first-out (FIFO) instructions are shift-register instructions that unload values in the same order that they were loaded. For example, they are often used

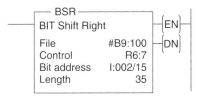

Figure 8.4 SLC 500 BSR instruction.

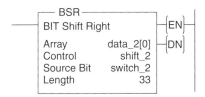

Figure 8.5 ControlLogix BSR instruction.

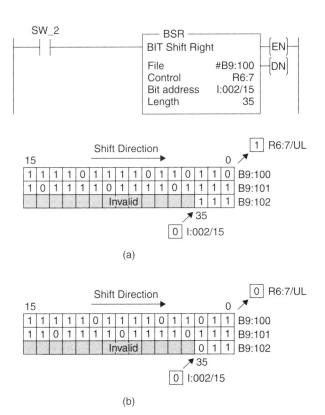

(a)

(b)

Figure 8.6 Operation of the BSR instruction; (a) Before instruction enabled; (b) After instruction enabled.

for tracking parts through an assembly line, where parts are represented by values that have a part number or an assembly code.

The FFL and FFU instructions are most often used in pairs to store and retrieve data in the order stored (first in, first out).

The SLC 500 FFL instruction is shown in Figure 8.7. The following parameters need to be entered in the SLC 500 FFL instruction:

Source	Word address from which the data are entered into the FIFO file.
FIFO	Address of the file in which the data are entered. The address must start with a # sign.
Control	R data-table type. The address is unique to the instruction. It is a three-word element that consists of the status word, the length, and the position.
Length	File length in words. The maximum value is 128 words.
Position	Next location from which data are entered when the instruction goes from false to true.

The status bits of the control word are the enable (en), the done (dn), and the empty (em) bits. The enable bit follows the instruction's status, the done bit is set when the instruction's position equals the length, and the empty bit is set when all the data have been unloaded from the FIFO file. When the done bit is set, the FIFO is full and does not accept any more data. When the done bit is set, the data in the FIFO file are not overwritten when the instruction goes from false to true.

Figure 8.8 shows the SLC 500 FFU instruction. The following parameters need to be entered in the SLC 500 FFU instruction:

FIFO	Address of the file in which the data are entered. The address must start with a # sign. When paired with an FFL instruction, this address is the same as the address for the FFL.
Destination	Address to which the FFU unloads data.
Control	R data-table type. The address is unique to the instruction. It is a three-word element that consists of the status word,

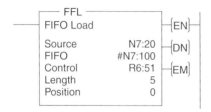

Figure 8.7 SLC 500 FIFO load (FFL) instruction.

the length, and the position. When it is paired with the FFL, the control addresses are the same.

Length   File length in words. The maximum value is 128 words.

Position   Next location from which data are unloaded when the instruction goes from false to true.

The status bits of the control word are the enable (en), the done (dn), and the empty (em) bits. The enable bit follows the instruction's status, the done bit is set when the instruction's position equals the length, and the empty bit is set when all the data have been unloaded from the FIFO file.

Figure 8.9 shows the FFL instruction for ControlLogix controllers. It operates in the same manner as the FFL instruction for SLC 500 controllers; they differ only in the data types used as the operands. The operands for the FFL in ControlLogix controllers are shown in Table 8.2.

Figure 8.10 shows the FFU instruction for ControlLogix controllers. It operates in the same manner as the FFU for SLC 500 controllers, differing only in the data types used as the operands. The operands for the FFU in the ControlLogix controllers are shown in Table 8.3.

Figure 8.11 shows an example using the SLC 500 FFL and FFU instructions. The ControlLogix FFL and FFU operate in the same way. When the FFL instruction is enabled, the en bit is set and the value at the source element is loaded into the next available element determined by the position. One element of data is loaded each time the instruction is enabled. When the FIFO file is full, the dn bit is set and no further data may be loaded.

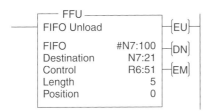

Figure 8.8   SLC 500 FFU instruction.

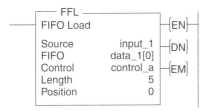

Figure 8.9   ControlLogix FFL instruction.

TABLE 8.2 FFL OPERANDS

Operand	Data	Format	Description
Source[*]	SINT INT DINT REAL structure	Tag immediate	Address of the data being shifted into the FIFO array
FIFO	SINT INT DINT REAL structure	Array tag	Address of the first element of the array where the data in the FIFO instruction is stored
Control	CONTROL	Tag	Control structure for the FIFO instruction; typically the control address for the FFL and FFU are the same
Length	DINT	Immediate	The length of the FIFO array; also the maximum number of values which can be entered into the FIFO array
Position	DINT	Immediate	Memory location where the next value is entered when the FFL is executed

[*]Source converts to the data type of the array tag. A smaller integer converts to a larger integer by sign conversion.

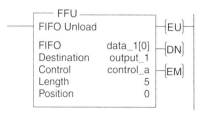

Figure 8.10 ControlLogix FFU Instruction.

When the FFU is enabled, the eu bit is set, and data from the first element of the FIFO file are unloaded into the destination element. One element of data is unloaded each time the instruction is enabled. When the FIFO file is empty, the em bit is set.

TABLE 8.3 FFU OPERANDS

Operand	Data	Format	Description
FIFO	SINT INT DINT REAL structure	Array tag	Address of the first element of the array where the data in the FIFO instruction is stored
Destination*	SINT INT DINT REAL structure	Tag	Address to which the FFU writes from the FIFO file when the unload instruction is enabled
Control	CONTROL	Tag	Control structure for the FIFO instruction; typically the control address for the FFL and FFU are the same
Length	DINT	Immediate	The length of the FIFO array; also the maximum number of values that can be entered into the FIFO array
Position	DINT	Immediate	Memory location from where the next value will be unloaded when the FFU is executed

* The destination value is converted to the data type of the destination. Smaller integers are converted to a larger integer by sign extension.

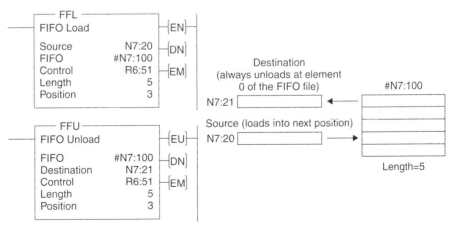

Figure 8.11 FIFO example.

8.1.4 LIFO Load (LFL) and LIFO Unload (LFU) Instructions

The last-in/first-out (LIFO) instructions are shift-register instructions that unload values in the reverse order that they were loaded. For example, they are often used for tracking parts on a stack, where they are tracked by a part number or an assembly code.

The LFL and LFU instructions are used in pairs to store and retrieve data in the order stored (last in, first out).

The LFL instruction for the SLC 500 controller is shown in Figure 8.12. The following parameters need to be entered in the SLC 500 LFL instruction:

Source Word address from which the data are entered into the LIFO file.

LIFO Address of the file in which the data are entered. The address must start with a # sign.

Control R data-table type. The address is unique to the instruction. It is a three-word element that consists of the status word, the length, and the position.

Length File length in words. The maximum value is 128 words.

Position Next location in the LIFO file where data are entered when the instruction goes from false to true.

The status bits of the control word are the enable (en), the done (dn), and the empty (em) bits. The enable bit follows the instruction's status, the done bit is set when the instruction's position equals its length, and the empty bit is set when all the data have been unloaded from the LIFO file. When the done bit is set, the LIFO is full and does not accept any more data. When the done bit is set, the data in the LIFO file are not overwritten when the instruction goes from false to true.

The LFU instruction for the SLC 500 controller is shown in Figure 8.13. The following parameters must be entered in the SLC 500 LFL instruction:

LIFO Address of the file in which the data are entered. The address must start with a # sign.

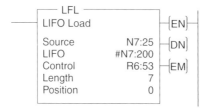

Figure 8.12 SLC 500 LFL instruction.

Destination	Address to which the value unloaded from the LFU instruction is written.
Control	R data-table type. The address is unique to the instruction. It is a three-word element that consists of the status word, the length, and the position. When paired with an LFL instruction, both have the same control address.
Length	File length in words. The maximum value is 128 words.
Position	Next location in the LIFO file where data are entered when the instruction goes from false to true.

The status bits of the control word are the enable (en), the done (dn), and the empty (em) bits. The enable bit follows the instruction's status, the done bit is set when the instruction's position equals its length, and the empty bit is set when all the data have been unloaded from the LIFO file.

Figure 8.14 shows the LFL instruction for ControlLogix controllers. It operates in the same manner as the LFL instruction for SLC 500 controllers. They differ only in the data types they use as operands. The operands for the LFL instruction in ControlLogix controllers are shown in Table 8.4.

Figure 8.15 shows the LFU instruction for ControlLogix controllers. It operates in the same manner as the LFU instruction for SLC 500 controllers. They differ only in the data types they use as operands. The operands for the LFU instruction in ControlLogix controllers are shown in Table 8.5.

The operation of LIFO instructions is illustrated in Figure 8.16. When the LFL instruction is enabled, the en bit is set and the value at the source element is

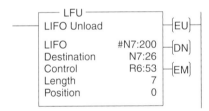

Figure 8.13   SLC 500 LFU instruction.

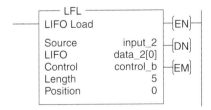

Figure 8.14   ControlLogix LFL instruction.

TABLE 8.4 LFL OPERANDS

Operand	Data	Format	Description
Source*	SINT INT DINT REAL structure	Tag immediate	Address of the data being shifted into the LIFO array
FIFO	SINT INT DINT REAL structure	Array tag	Address of the first element of the array where the data in the LIFO instruction are stored
Control	CONTROL	Tag	Control structure for the LIFO instruction; typically the control address for the LFL and LFU are the same
Length	DINT	Immediate	Length of the LIFO array; also the maximum number of values that can be entered into the LIFO array
Position	DINT	Immediate	Memory location where the next value is entered when the LFL instruction is enabled

*Source converts to the data type of the array tag. A smaller integer converts to a larger integer by sign conversion.

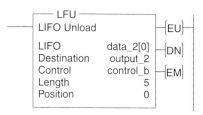

Figure 8.15 ControlLogix LFU instruction.

loaded into the next available element, as determined by the position. One element of data is loaded each time the instruction is enabled. When the LIFO file is full, the dn bit is set and no further data may be loaded.

When the LFU is enabled, the en bit is set, and data are unloaded from the position last loaded into the destination element. One element of data is unloaded each time the instruction is enabled. When the LIFO file is empty, the em bit is set.

TABLE 8.5 LFU OPERANDS

Operand	Data	Format	Description
LIFO	SINT INT DINT REAL structure	Array tag	Address of the first element of the array where the data in the LIFO instruction is stored
Destination*	SINT INT DINT REAL structure	Tag	Address to which the LFU writes from the LIFO file when the unload instruction is enabled
Control	CONTROL	Tag	Control structure for the LIFO instruction; typically the control address for the LFL and LFU are the same
Length	DINT	Immediate	Length of the LIFO array; also the maximum number of values that can be entered into the LIFO array
Position	DINT	Immediate	Memory location from where the next unload takes place

*The destination value is converted to the data type of the destination. Smaller integers are converted to a larger integer by sign extension.

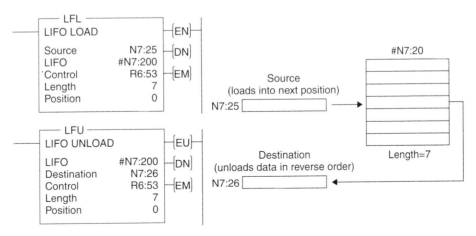

Figure 8.16   LIFO example

▶ 8.2 SEQUENCER INSTRUCTIONS

Sequencer instructions are used to control sequential process operations, where a repeated sequence is executed. The sequencer instructions differ for SLC 500 and ControlLogix controllers. The SLC 500 controller has sequencer compare, sequencer output, and sequencer load instructions. The ControlLogix controller has sequencer input, sequencer output, and sequencer load instructions.

8.2.1 SLC 500 Sequencer Instructions

The SLC 500 controller has three sequencer instructions: the sequencer compare (SQC) instruction, the sequencer output (SQO) instruction, and the sequencer load (SQL) instruction.

The SQC instruction is illustrated in Figure 8.17. The following parameters need to be entered in the SLC 500 SQC instruction:

File	Address of the file in which the compare data are entered. The address must start with a # sign. The data in this file are compared to the data in the source.
Mask	Hexadecimal value, word address, or file address through which the source value is masked when compared to the file address. If the mask is a file address, it will track directly with the file address. The mask may be directly entered in binary, decimal, or hexadecimal, but the display is always in hexadecimal.
Source	Address to which the current step's data in the file are compared through the mask.
Control	R data-table type. The address is unique to the instruction. It is a three-word element that consists of the status word, the length, and the position.
Length	Number of steps, starting with position 1. This means the actual file length will be one word longer than the length indicated in the instruction.

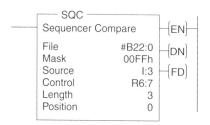

Figure 8.17 Sequencer compare instruction.

The status bits of the control word are the enable (en), the done (dn), the error (er), and the found (fd) bits. The enable bit follows the instruction's status, the done bit is set when the position is equal to the length, and the found bit is set when there is a match between the filtered data from the source to the comparative bits in the file at the position indicated. The error bit is set when the position value is negative.

Figure 8.18 shows an example of an SQC instruction. The SQC indexes its position by 1 every time SW_1 is toggled. If it starts on position 0, the next time the switch is toggled it will index to position 1. This continues until the instruction is on Step 3. When the position is 3 and SW_1 is toggled again, the position indexes to position 1. The only way to return to position 0 is to move a zero into the position address (R6:7.POS) or use the reset instruction with the address of the control (R6:7). Position zero is used as the starting point, or the home position.

Because the position is 2, the data in the lower 8 bits of the source (I:3) are compared to the data in the file at Step 2 (B22:2). The lower 8 bits are compared based on the status of the bits in the mask. In the example, the mask is 00FF in hexadecimal. Because the lower 8 bits in I:3 match the status of the lower 8 bits in B22:2, the found (fd) bit is set. The fd bit may then be used to control devices based on the comparison.

The SQO instruction is illustrated in Figure 8.19. The following parameters need to be entered in the SLC 500 SQO instruction:

File   Address of the file from which the data are copied through the mask to the destination. The data must be entered in this file to properly turn the destination bits ON and OFF in the intended sequence.

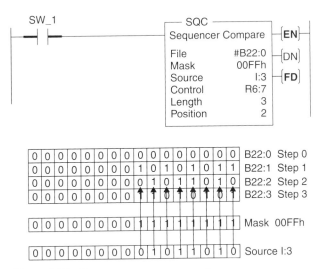

Figure 8.18 Sequencer compare example.

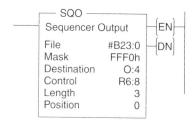

Figure 8.19 Sequencer output.

Mask	Hexadecimal value, word address, or file address through which the file value is copied to the destination address. If the mask is a file address, it will track directly with the file address. The mask may be entered directly in binary, decimal, or hexadecimal, but the display is always in hexadecimal.
Control	R data-table type. The address is unique to the instruction. It is a three-word element that consists of the status word, the length, and the position.
Length	Number of steps, starting with position 1. The actual file length is one word longer than the length indicated in the instruction.

The status bits of the control word are the enable (en), the done (dn), and the error (er) bits. The enable bit follows the instruction's status, and the done bit is set when the position is equal to the length. The error bit is set when the position value is negative.

Figure 8.20 shows an example of an SQO instruction. The SQO will index its position by 1 every time SW_1 is toggled. If it starts on position 0, the next time the switch is toggled it will index to position 1. This continues until the instruction is on Step 3. When the position is 3 and SW_1 is toggled again, the position indexes to position 1. The only way to return to position 0 is to move a 0 into the position address (R6:8.POS) or use the Reset instruction with the address of the control (R6:8). Position 0 is used as the starting point, or the home position.

Because the position is 2, the data in the upper 12 bits of the file address, B23:2, are copied through the mask to the destination word, O:4. In the example, the mask is FFF0 in hexadecimal notation. This blocks the lower 4 bits from being copied; thus they remain in their last state and are not affected by the operation.

The sequencer load (SQL) instruction is shown in Figure 8.21. The SQL instruction is normally used to load data into the SQC's file. The machine is stepped manually through its cycle, and at each position the inputs are read and copied into the appropriate file address position.

The following parameters must be entered in the SLC 500 SQL instruction:

File	Address of the file into which data are copied from the source. All bits are copied, because there is not a mask.

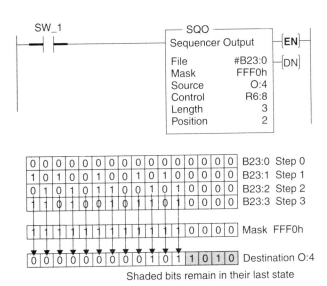

Figure 8.20   Sequencer output example.

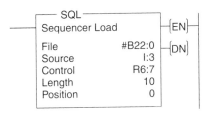

Figure 8.21   Sequencer load instruction.

Source   Word address from which data are copied into the position of the file, as indicated in the instruction.

Control   R data-table type. The address is unique to the instruction. It is a three-word element that consists of the status word, the length, and the position.

Length   Number of steps, starting with position 1. The actual file length is 1 word longer than the length indicated in the instruction.

The status bits of the control word are the enable (en), the done (dn), and the error (er) bits. The enable bit follows the instruction's status, and the done bit is set when the position is equal to the length. The error bit is set when the position value is negative.

Figure 8.22 shows an example of the sequencer load operation. Each time SW_3 goes from false to true, the SQL instruction indexes one position and copies the data from the source into the current position in the file. In the current position,

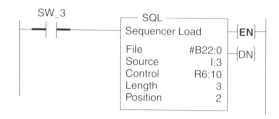

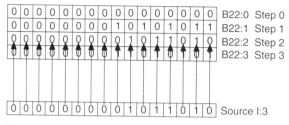

Figure 8.22 Sequencer load example.

the data in I:3 are copied to B22:2, which is Step 2. The SQL is used to load the data in the SQL file by manually stepping the machine through its sequence and reading the inputs at each step. Otherwise, the data would have to be entered into the file manually.

8.2.2 ControlLogix Sequencer Instructions

The SQI instruction is shown in Figure 8.23. The SQI operates in a manner similar to a masked equal instruction. The difference is that the source value is compared to the value at the current position in the array. If the two values are equal, the instruction is logically true. Otherwise, it is logically false. The SQI cannot manipulate its own position. It must be externally indexed, such as by an SQO instruction with the same control address.

Table 8.6 shows the operands for the SQI instruction.

In Figure 8.24, the instruction is using the parameters from the instruction in Figure 8.23. The first 8 bits of the source value are compared to the first 8 bits in data[2]. This is because the mask is the hexadecimal value 00FF. In this example, the first 8 bits of the source value do match the first 8 bits in data[2], so the instruction is logically true. The SQI can be used with other input instructions in the rung.

Figure 8.25 shows the SQO instruction. When the SQO is enabled, the position is incremented by 1, and the data in the array at that position are copied through the mask to the destination address.

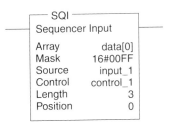

```
┌─ SQI ──────────────┐
│ Sequencer Input    │
│                    │
│ Array       data[0]│
│ Mask        16#00FF│
│ Source      input_1│
│ Control   control_1│
│ Length            3│
│ Position          0│
└────────────────────┘
```

Figure 8.23   ControlLogix sequencer input (SQI) instruction.

TABLE 8.6   SQI OPERANDS

Operand	Data	Format	Description
Array	DINT	Array tag	First element of the sequencer array
Mask	SINT[*] INT[*] DINT	Tag immediate	Determines which bits to compare or ignore
Source	SINT[*] INT[*] DINT	Tag	Input value compared to value
Control	CONTROL	Tag	Control structure for the SQI instruction; typically the control address for the SQI and SQO are the same
Length	DINT	Immediate	Length of the SQI array
Position	DINT	Immediate	Current position of the array

[*]A SINT or INT tag converts to a DINT value by sign extension.

Example values (DINTs displayed in binary)

Array	00000000 00000000 00000000 10101010	data[2]
Mask	00000000 00000000 00000000 11111111	00FF
Source	00000000 00000000 00000000 10101010	input_1

Shaded bits are not compared due to the zeros in the mask.

Position=2

Figure 8.24   Sequencer input example.

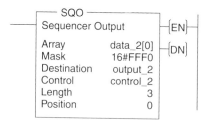

Figure 8.25 Sequencer output (SQO) instruction.

TABLE 8.7 SQO OPERANDS

Operand	Data	Format	Description
Array	DINT	Array tag	First element of the sequencer array
Mask	SINT* INT* DINT	Tag immediate	Determines which bits to block or pass
Destination	DINT	Tag	Output data from the sequencer array
Control	CONTROL	Tag	Control structure for the SQO instruction
Length	DINT	Immediate	Length of the SQO array
Position	DINT	Immediate	Current position of the array

*A SINT or INT tag converts to a DINT value by sign extension.

The operands for the SQO are shown in Table 8.7. The control structure contains the status bits, the length, and the position. The status bits are the enable (en), the done bit (dn), and the error (er) bit. The enable bit follows the condition of the SQO, and the done bit is set when all the elements of the array have been copied to the destination.

In Figure 8.26, the SQO is indexed to Position 2. The data in the upper 12 bits of Step 2 of the array are then copied through the mask to the upper 12 bits of the destination.

Figure 8.27 shows an SQI instruction being used with an SQO instruction. When used this way, the control, length, and position must be the same for both instructions. When the SQI is true at the position it is in, it enables the SQO; the SQO then indexes and copies data from the array position to the destination. Thus, the SQI going true causes the SQO to execute the next position.

The SQL instruction is shown in Figure 8.28. The SQL instruction is used to load data into a file. Instead of going into the array in the software and entering the

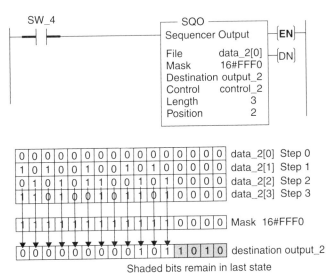

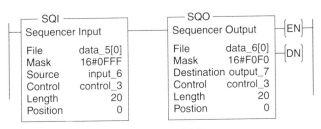

Shaded bits remain in last state

Figure 8.26 SQO example.

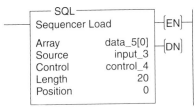

Figure 8.27 Using an SQI with an SQO.

Figure 8.28 Sequencer load (SQL) instruction.

data, you move the machine manually through the sequence. At each step in the sequence, the source (input) data are copied into that position in the array. The SQL performs an element-to-file copy in the incremental mode.

The SQL instruction requires the parameters shown in Table 8.8.

Figure 8.29 shows the operation of the SQL instruction. In this example the position is indexed to 2, and the data in input_3 are copied to Step 2 in the array (data_5[2]). The SQL is used to load the data in the SQL file by manually stepping the machine through its sequence and reading the inputs at each step. Otherwise, the data would have to be manually entered into the file.

TABLE 8.8 SQL OPERANDS

Operand	Data	Format	Description
Array	DINT	Array tag	First element of the sequencer array
Source	SINT	Tag	Source data to be copied
	INT	Immediate	
	DINT		
Control	CONTROL	Tag	Control structure for the SQL instruction
Length	DINT	Immediate	Length of the SQL array
Position	DINT	Immediate	Current position of the array

*A SINT or INT tag converts to a DINT value by sign extension.

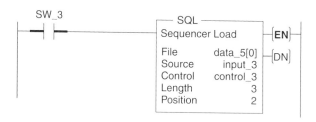

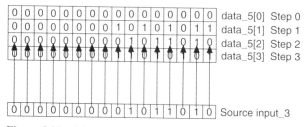

Figure 8.29 SQL example.

SUMMARY

The BSL, BSR, FFL, FFU, LFL, and LFU instructions are commonly used to track parts. The bit-shift instructions track by bit status, so we track 0s and 1s, which can represent tracking part present/part missing or good part/bad part. The bit-shift instruction is indexed each time the machine is indexed so that they remain in sync. The FIFO and LIFO instructions track on a first-in/first-out or last-in/last-out basis. Tracking is done on a word level, so part numbers, colors, sizes, etc., may be tracked.

► EXERCISES

1. What instruction could be used to monitor whether or not a part is present on a 23-station conveyor line?

2. What instruction could be used to control a pumping operation that repeats the same sequence of operation?

3. What instructions could track parts in a holding area by part number? The parts are to be removed from the holding area in the same order that they were put in.

4. Design the ladder logic that will monitor whether there are parts present or not present on a ten-station conveyor. Show the logic that will turn on an output if there is a part present at Station 5 and another output if there is a part present at Station 7. Also turn on an output if there is no part at Station 3. The stations are numbered 1–10. SW_5 senses when a part is loaded in Station 1. The conveyor is to index every 2 s.

5. Add the logic to Exercise 4 that would allow for the tracking of parts on the same line if there were a problem with the line and the line had to be reversed to get it back into sequence. Manually reverse with switch SW_6. The input to the shift register is switch SW_7.

6. Design the ladder logic that will monitor parts by part number as parts are placed in a holding area. The parts are to be removed from the holding area in the same order they were put in. The holding area can hold ten parts. Part numbers being loaded are 1234, 3456, and 8765. Turn on an output if part 1234 is unloaded. Turn on a second output if 3456 is unloaded, and turn on a third output if 8765 is unloaded. Switch SW_10 loads data and switch SW_11 unloads data.

7. Design the ladder logic that will monitor parts being placed on a stack. The parts will be unloaded in the reverse order that they were placed on the stack. Fifteen parts can be loaded on the stack. Part numbers being loaded are 9876, 5677, and 5665. Turn on an output if 9876 is unloaded, a second output if 5677 is unloaded, and a third output if 5665 is unloaded. Switch SW_12 will load data and switch SW_13 will unload data.

8. Design the ladder logic for an SLC 500 controller that will control the first 8 bits in B11:0, as shown in following sequence of operation. The outputs will sequence every 5 s.

    Step 0: all bits off

    Step 1: bits 0, 2, 6, and 7 on; all others off

    Step 2: bits 1, 3, 4, and 6 on; all others off

    Step 3: bits 2, 3, 4, and 5 on; all others off

    Step 4: bits 1 and 6 on; all others off

    Step 5: bits 2, 3, 5, and 7 on; all others off

9. For an SLC 500 controller, set up an SQC instruction that will set a bit whenever a match is found, as shown in the following chart. The SQC will monitor the first 9 bits in B35:0. The SQI will index every time SW_6 transitions from off to on.

    Step 0: all bits off

    Step 1: bits 0, 2, 5, 7, and 8 on; all others off

    Step 2: bits 1, 3, and 4 are on; all others off

    Step 3: bits 3, 5, and 6 on; all others off

    Step 4: bits 1, 5, and 7 on; all others off

10. Design the ladder logic for a ControlLogix controller using sequencer instructions that will monitor the status of inputs and control outputs as shown in the following chart. Monitor the first 9 bits of the inputs and control the first 7 bits of the outputs; sw_1 manually indexes the outputs through their steps.

Inputs to be monitored (tag: status_1):

    Step 0: all bits off

    Step 1: bits 0, 3, 5, and 8 on; all other bits off

    Step 2: bits 1, 4, and 5 on; all other bits off

    Step 3: bits 2, 3, 6, and 7 on; all other bits off

    Step 4: bits 1, 4, and 5 on; all other bits off

    Step 5: bits 3, 6, 7, and 8 on; all other bits off

Outputs to be controlled (tag: output_1):

    Step 0: all bits off

    Step 1: bits 0, 5, and 6 on; all other bits off

    Step 2: bits 1, 2, 3, and 5 on; all other bits off

    Step 3: bits 2, 4, and 5 on; all other bits off

    Step 4: bits 0, 1, and 6 on; all other bits off

    Step 5: bits 2 and 6 on; all other bits off

9

SLC 500 and ControlLogix Program-Control Instructions

CHAPTER OUTLINE

9.1 Master Control Reset (MCR) Instruction
9.2 Jump (JMP) and Label (LBL) Instructions
9.3 Subroutines
9.4 Selectable Timed Interrupt File
9.5 SLC 500 Immediate Input with Mask (IIM) and Immediate Output with Mask (IOM) Instructions

---

LEARNING OBJECTIVES

Upon reading this chapter students should be able to:

- Explain program-control concepts.
- Interpret and program the Master Control Reset (MCR) instruction.
- Interpret and program the Jump (JMP) and Label (LBL) instructions.
- Interpret and program subroutines.
- Interpret and program Immediate Input with Mask (IIM) and the Immediate Output with Mask (IOM) instructions.

INTRODUCTION

This chapter covers the program-control instructions that change the way a program is scanned. The differences between the SLC 500 and ControlLogix instructions and the way programs are scanned are explained in this chapter.

The SLC 500 controller uses program file 2 as the main program. Files 3–255, if they are used, are assigned as **subroutine** files. The ControlLogix controller has a project that can contain tasks, programs, and routines, as discussed in Chapter 3.

The program-control instructions covered in this chapter include the MCR, JMP, LBL, JSR, SBR, RET, IIM, and IOM instructions. Also, the Selectable Timed Interrupt (STI), which is a specialized routine, is covered.

▶ 9.1 MASTER CONTROL RESET (MCR) INSTRUCTION

The Master Control Reset (MCR) instruction operates the same in the SLC 500 controller as it does in the ControlLogix controller. You use the MCR instruction in pairs, which creates a zone between the instructions. It does not have an address. The instruction allows all rungs in the zone to function as false rungs. You program the first MCR with input instructions in the rung and the second MCR without any other instructions in the rung. You cannot nest the zones. If you create a zone at the end of the program or routine, it is not necessary to program an MCR, because it will treat the end of the program or routine as the end of the zone. You may create multiple MCR zones within a program or routine.

In relay logic MCR may refer to a Master Control Relay, which is a hardwired device that provides capability for emergency stops. The MCR in the controller is *not* a substitute for a Master Control Relay. All emergency stops are hardwired.

Figure 9.1 shows an MCR zone between the first and last rungs. In this situation, the first MCR is true. When the first MCR is true, the rungs between the MCRs function normally. All rungs within the zone are scanned and evaluated for true or false conditions. Because SW_4 and SW_6 are ON, Output_1 will be latched and Output_2 will be ON.

In Figure 9.2, SW_1 and SW_4 are turned off. Because the first MCR is false, all rungs between the two MCRs are acted on as though they are false. A false rung

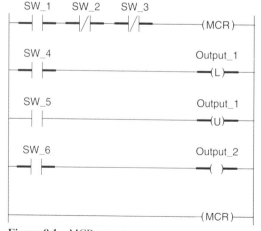

Figure 9.1 MCR zone true.

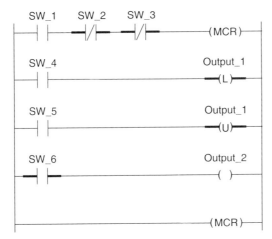

Figure 9.2 MCR zone false.

leaves the latch in its last state, and the output energize is turned off. When troubleshooting a project, you need to know whether rungs are within zones. As you can see, you might be confused to see SW_6 true and Output_2 false, unless you knew they were inside a zone.

▶ 9.2 JUMP (JMP) AND LABEL (LBL) INSTRUCTIONS

You can use Jump (JMP) and Label (LBL) instructions to jump over a section of program that you do not want to scan at that time. The JMP is an output instruction, and the LBL is an input instruction. The JMP jumps to the LBL with the same identifier. The LBL to which the JMP jumps must be in the same program file or routine, and the LBL instruction must be the first instruction in the rung. You can have multiple JMPs with the same identifier in a program file or routine, but you can have only one LBL with that identifier. In an SLC 500 controller the JMP will have a number from 0 to 255, and the corresponding LBL will have the same number. There can be up to 256 labels in each file, but the total number of labels in the project cannot exceed 256. In the ControlLogix controller the JMP and corresponding LBL have the same name. The name can be up to 40 characters, including letters, numbers, and underscores.

In Figure 9.3, the JMP instruction is false, so all the rungs between the JMP and the LBL are scanned in their normal fashion and the rungs execute based on their input conditions. The latch is set, and Output_2 and Output_3 are turned ON.

In Figure 9.4, the JMP to LBL is true, so the rungs between the JMP and the LBL are not scanned. In this example, SW_1 is turned ON, SW_4 is turned OFF, SW_6 is turned OFF, and SW_7 is turned off, in that order. As shown, the outputs—

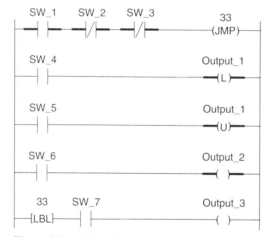

Figure 9.3 JMP to LBL false.

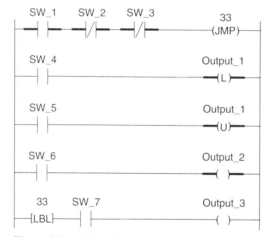

Figure 9.4 JMP to LBL true.

both the latch and the output energize—remain in their last state. All outputs remain in their last state when they are not scanned. Output_3 is outside the zone, so it is not affected by the JMP to LBL. The LBL instruction is just a target for the JMP and does not affect the logic in the rung in which it is located.

You may also locate the JMP instruction after its corresponding LBL in the program. This causes repeat iterations of logic. You must be careful not to do an excessive amount of iterations because the watchdog timer could time out, causing the controller to fault.

▶ 9.3 SUBROUTINES

9.3.1 SLC 500 Subroutines

Subroutines in the SLC 500 are located in program files 3–255. There are three subroutine instructions that allow these program files to be scanned.

Figure 9.5 shows the three subroutine-related instructions. The JSR instruction causes the scan to jump to the program file designated in the instruction. It is the only parameter entered in the instruction.

The SBR instruction is the first instruction on the first rung in the subroutine file. It serves as an identifier that the program file is a subroutine. It is always true and is an optional instruction.

The RET instruction is an output instruction that causes the scan to return to the next instruction after the JSR in the program from which it jumped. The scan returns from the end of the file if there is no RET instruction. The RET instruction may also be conditional, depending on whether or not you want to scan rungs in the subroutine after the RET instruction.

Figure 9.6 shows an example of jumping to a subroutine and returning to the main program file. When the JSR instruction is true, the scan jumps to the first instruction in the first rung of the file designated in the JSR. It then scans the subroutine until it comes to a true RET instruction or the end of the file. The scan then returns to the instruction after the JSR and continues scanning that file.

Nested subroutines are shown in Figure 9.7. When the JSR is true in Program File 2, the scan jumps to the first instruction in the first rung in Program File 3. It then scans Program File 3; if the JSR instruction in File 3 is true, the scan jumps to Program File 5. It scans File 5 until it reaches the RET instruction and then it returns to the instruction after the JSR in File 3. It scans File 3 until it reaches the RET, returns to the instruction after the JSR in Program File 2, and scans File 2 to the end of the file.

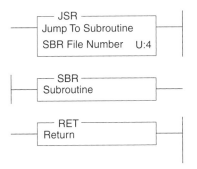

Figure 9.5 SLC 500 JSR, SUB, and RET instructions.

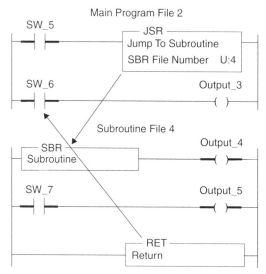

Figure 9.6 SLC 500 subroutine example.

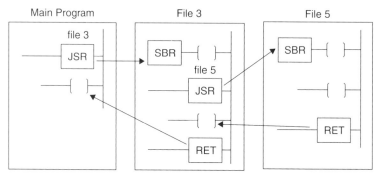

Figure 9.7 SLC 500 nested subroutines.

In fixed and 5/01 processors, you can nest subroutines up to four levels. In 5/02, 5/03, 5/04, and 5/05 processors, you can nest subroutines up to eight levels.

9.3.2 ControlLogix Subroutines

Three instructions are associated with jumping to subroutines. They are the *Jump to Subroutine (JSR)*, the *Subroutine (SBR)*, and the *Return (RET)* instructions, shown in Figure 9.8.

The JSR is an output instruction. A true JSR instruction causes the scan to jump to the named subroutine in JSR. The SBR is an input instruction and is the first instruction in the subroutine. It is used as an identifier only if parameters are

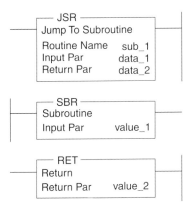

Figure 9.8 ControlLogix JSR, SBR, and RET Instructions.

not being passed to the subroutine. It is also always true. The RET returns the scan to the instruction after the JSR. If there is no RET instruction in the subroutine, the end-of-file returns the scan.

Passing parameters allows for data to be copied to and from the subroutine using the input parameters and the return parameters. The value(s) stored in the input parameter(s) in the JSR are copied to the input parameter(s) in the SBR. The value(s) stored in the return parameter(s) in the RET are copied back to the return parameter(s) in the JSR. The number of input parameters in the JSR must match the number of input parameters in the SBR, and the number of return parameters in the RET must match the number of return parameters in the JSR. The value stored in the tag in the first input parameter in the JSR is copied to the first input parameter in the SBR, etc. The same is true for the return parameters. For each input parameter operand in the JSR, the corresponding input parameter in SBR must be the same data type, including any array dimensions. This is also true for the corresponding return parameters in the JSR and the RET.

It is not necessary to pass parameters when jumping to a subroutine. Just leave the input parameter and return parameter blank if you do not wish to pass parameters.

When passing parameters, as shown in Figure 9.9, the value stored at the tag address in the JSR input parameter data_1 is copied to the tag address value_1 in the SBR input parameter when the scan jumps to the subroutine sub_1. When the scan returns to the main routine, the value stored at the tag address value_2 is copied to the tag address data_2 in the return parameter in the JSR.

Passing parameters is often used when the subroutine is used to do calculations on data. Different data values may be copied to the subroutine, and the results may be copied back to the main program. There are no limits, other than processor memory, on the number of parameters that can be passed or returned.

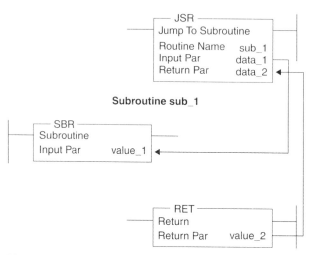

Figure 9.9 Jump to subroutine with passing parameters.

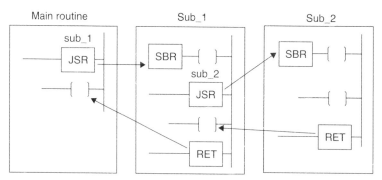

Figure 9.10 ControlLogix nested subroutines.

Subroutines can also be nested, where one subroutine is called from another subroutine. This is shown in Figure 9.10. There is no limit to the number of nested subroutines.

The JSR in the main routine causes the scan to jump to sub_1. It then scans sub_1 until it comes to the true JSR; from there it jumps to sub_2. It then scans sub_2 until it reaches the RET. From there it returns to the instruction after the JSR in sub_1 and continues scanning sub_1 until it reaches the RET instruction. From this point it returns to the instruction after the JSR in the main routine. If the JSR in sub_1 is not true, the scan scans sub_1 to the RET and returns to the instruction after the JSR in the main routine.

► 9.4 SELECTABLE TIMED INTERRUPT FILE

As discussed in Chapter 3, the project memory of the ControlLogix controller is set up as tasks, programs, and routines. The tasks may be configured as continuous or periodic tasks. There can be 1 continuous task and 31 periodic tasks; if there is not a continuous task, there can be 32 periodic tasks. The periodic tasks are triggered at a repeated time interval, from 1 ms to 2000 ms, with the default time being 10 ms. Tasks also have priorities from 1 through 15, with 1 being the highest priority. A higher-priority task interrupts a lower-priority task. The continuous task always has the lowest priority. The periodic tasks are similar to the selectable timed interrupt program files in an SLC 500 controller.

In the SLC 500, program files 3–255 are used for subroutines. One of these files may be assigned as a selectable timed interrupt file.

Figure 9.11 shows the setup of the STI in the processor status area of memory. Status word S:30 stores the set point of the STI. This is the time period at which the program scan is interrupted and the STI file is executed. Status word S:31 stores the STI program file number. Both the set point and the file may be manipulated from the program logic, although there can only be one STI file at any given time.

The PID instruction is often placed in the STI, because it needs to be executed at a precise time interval. Scan time can vary in a program, which means that an instruction in a regular portion of the program may not execute at a repeated time interval. Placing the PID instruction in the STI ensures that it executes at the same repeated time interval.

You can place Immediate Input with Mask and Immediate Output with Mask instructions in an STI file in high-speed applications. This allows the scan to be interrupted, so inputs and outputs can be updated at a predictable time interval and at a higher rate than once per scan.

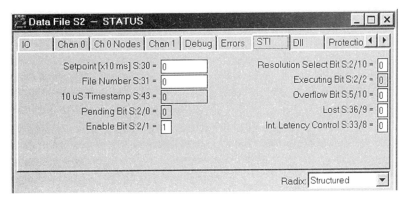

Figure 9.11   Selectable timed interrupt setup.

▶ 9.5 SLC 500 IMMEDIATE INPUT WITH MASK (IIM) AND IMMEDIATE OUTPUT WITH MASK (IOM) INSTRUCTIONS

The Immediate Input with Mask (IIM) and Immediate Output with Mask (IOM) instructions are only in the SLC 500 controllers. They are used in high-speed applications where the input transition may not be seen due to its short transition duration compared to the program scan time and where the output information needs to be sent to the real-world output before the end of the program scan. In ControlLogix controllers, the I/O is updated asynchronously to the program scan, and output information is sent at the end of the rung, so the IIM- and IOM-type instructions are not needed.

It is common to put the IIM and IOM instructions in an STI. That way they are executed at a controlled rate, as opposed to being in other program files. There, their execution depends on the program's scan time, which is not consistent.

The IIM instruction interrupts the program scan to update input data from the module located in the slot specified in the instruction. The data are transferred through the mask to the input data table. Data are transferred at the bit locations where there is a 1 in the mask. Data are not transferred where there is a 0 in the mask.

The IIM instructions for the controllers indicated are shown in Figure 9.12. The following parameters are entered in the instruction:

Slot	The slot and word that contain the data to be updated. In the instruction shown, it is slot 3, word 0. In the fixed and 5/01 controllers, there can be up to 8 words (0–7) associated with the slot. In the 5/02, 5/03, 5/04, and 5/05 controllers there can be up to 32 words (0–31) associated with the slot.
Mask	Either an address or a constant. If you enter a constant you may enter it in binary, decimal, or hexadecimal. The software

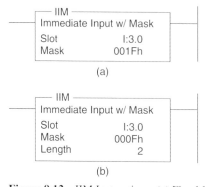

Figure 9.12   IIM Instructions; (a) Fixed SLC 5/01 and 5/02; (b) SLC 5/03, 5/04, and 5/05.

converts the display to hexadecimal. A 1 in the mask allows the data to be updated and a 0 blocks.

Length   (5/03, 5/04, and 5/05 only) Used to transfer more than one word per slot.

The IOM instruction interrupts the program scan to update output data to the module located in the slot specified in the instruction. The data are transferred through the mask to the output module. Data are transferred at the bit locations where there is a 1 in the mask. Data are not transferred where there is a 0 in the mask.

The IOM instructions for the controllers indicated are shown in Figure 9.13. The following parameters are entered in the instruction:

Slot   The slot and word that contain the data to be updated. In the instruction shown, it is slot 4, word 0. In the fixed and 5/01 controllers, up to 8 words (0–7) can be associated with the slot. In the 5/02, 5/03, 5/04, and 5/05 controllers, up to 32 words (0–31) can be associated with the slot.

Mask   Either an address or a constant. If you enter a constant, you may enter it in binary, decimal, or hexadecimal. The software will convert the display to hexadecimal. A 1 in the mask allows the data to be updated and a 0 blocks.

Length   (5/03, 5/04, and 5/05 only) Used to transfer more than one word per slot.

Figure 9.14 shows an example of the use of the IIM and IOM instructions. It is desirable to have this logic located in an STI file so the rungs can be executed at a controlled time rate. In the first rung, the status of terminal 3 on the input module located in slot 3 is updated into the input data table when the rung is scanned. Due to value in the mask, no other inputs from this module are updated. If this

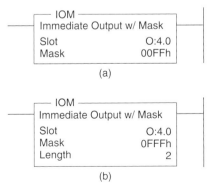

(a)

(b)

Figure 9.13   IOM instructions; (a) Fixed, SLC 5/01 and 5/02; (b) SLC 5/03, 5/04, and 5/05.

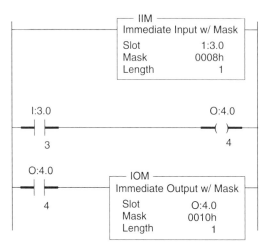

Figure 9.14 IIM and IOM example.

input is ON, the second rung is true, turning on bit 4 in word O:4.0. In the third rung, because bit O:4.0/4 is true, the IOM instruction sends the status of this bit to terminal 4 on the output module in slot 4. Due to the value in the mask, no other output terminals are updated at this time. All data in the input and output data table are updated during the usual portion of the scan.

This application can be used in a high-speed canning process, where the input is used to sense a defective can. When a defective can is detected, an air blast from an air valve blows the defective can off the line. The only time we need to energize the output and the IOM is when there is a defective can. That is why the IOM is conditional. However, we need to sense every can, so the IIM is unconditional.

SUMMARY

Routines are normally scanned from left to right and top to bottom. The program-control instructions alter the way the routine is normally scanned. In the SLC 500 there is a main program file, and there also may be subroutines. In the ControlLogix controller, the routines function similar to the main program and subroutines in the SLC 500.

The MCR is used within a main program or routine. MCRs come in pairs to create an MCR zone that starts with a conditional rung containing an MCR and ends with a rung containing an unconditional MCR. When the start rung is true, all the rungs within the zone are scanned and the logic functions are based on rung conditions. When the start rung is false, all the rungs in the zone are scanned as being false.

Another type of zone is the jump-to-label zone. The JMP instruction is an output instruction that, when true, jumps to an LBL instruction with the same identifier. The outputs on all the rungs not being scanned remain in their last state. You can jump to a label from more than one JMP and you can also jump to a previous rung.

The JSR, SBR, and RET instructions are used for jumping between routines and in the ControlLogix controller for passing parameters when jumping between the routines.

The Selectable Timed Interrupt (STI) is used to jump to a specified routine on a time basis where the scan will be interrupted at a specified timer interval and the STI routine will be executed. It is used only in the SLC 500 controller.

The IIM and IOM are available only in the SLC 500 controllers. They are used in high-speed applications where the scan is interrupted to update up to one slot of input or output data.

▶   EXERCISES

 1. What defines an MCR zone?
 2. What is the purpose of an MCR zone?
 3. What happens to the outputs in an MCR zone when the start rung is false?
 4. What happens to the outputs in an MCR zone when the start rung is true?
 5. What happens to the outputs in each rung between the MCRs in Figure 9.15 when SW_5 is turned off?
 6. What is the purpose of a jump-to-label section of logic?
 7. What happens to the outputs inside of a jump-to-label section of ladder when the section is jumped over?
 8. What is the function of an STI in an SLC 500 controller?
 9. What happens to the outputs in the section of ladder logic shown in Figure 9.16 when SW_8 is turned on, SW_4 is turned off, SW_6 is turned off, SW_9 is turned off, and SW_7 is turned off, in that order?
10. What is the function of a subroutine?
11. What is a nested subroutine?
12. How many levels of nested subroutines are allowed in the SLC 500 and ControlLogix controllers?
13. What is the function of passing parameters when using subroutines?
14. Design the logic for an SLC 500 controller that will immediately update terminal 4 of an input module located in slot 4 and—when that input is true—will immediately update terminal 9 of an output module located in slot 5.

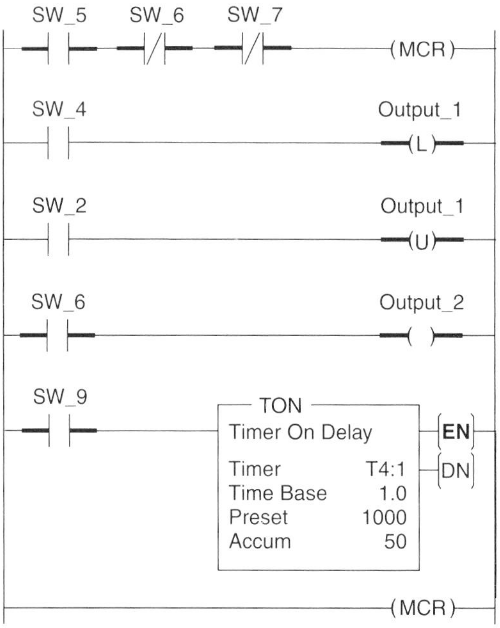

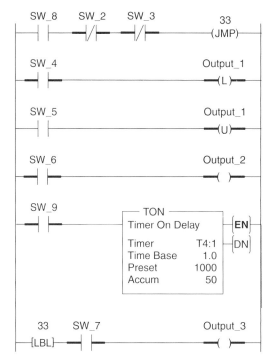

Figure 9.15 Exercise 5.

Figure 9.16 Exercise 9.

Structured Programming on the PLC

CHAPTER OUTLINE

LEARNING OBJECTIVES

Upon reading this chapter students should be able to:

- Generate a state diagram when given a control problem.
- Implement a control scheme using a state diagram and zone control.
- Implement a control scheme using a state diagram and subroutines.
- Design a sequential function chart when given a control problem.
- Give the advantages of structured versus nonstructured programming.

INTRODUCTION

There are several disadvantages to creating complex programs by trial and error. This method makes programs difficult to follow, to troubleshoot, and to document. It is also very inefficient to create and thoroughly test a program in this manner. This chapter introduces structured programming, which makes control problems more efficient to solve as well as easier to troubleshoot and document. You will learn how to apply state diagrams and Petri networks to organize your programming. These techniques use graphics to display the organization of control, which is invaluable documentation. You can use structure to break up complex control into several smaller tasks, which can be assigned to more than one person. Thus structured programming lends itself to a team approach rather than an individual one.

► 10.1 SEQUENTIAL CONTROL AND STATE DIAGRAMS

A powerful technique for solving complex control problems is the use of graphics, and a state diagram provides a means of doing this. A state diagram is basically a flowchart that shows the order and control of a process. One of the rules that must be followed is that no more than one state can be active at a time—that only one operation occurs at a time. You cannot proceed to the next step until you finish the previous step. The following example shows how a state diagram is created.

10.1.1 Example: Creating a State Diagram

A programmable controller controls a printed circuit board (PCB) etching process. The PCB is manually placed on an overhead conveyor, and the board travels on it to the first etching tank. When it is directly over the etching tank, the board is lowered into the etching tank by energizing a lowering motor. A limit switch on this mechanism provides a signal to shut off the motor. The board remains in the etching solution for 5 min and then is removed. The mechanism has a limit switch for signaling when the board has been raised to the proper height so that it can move to the next process. The board then continues to a rinse tank and is rinsed for 1 min. Finally, the board is put in a cleaning solution for 2 min, and then it goes to an inspection station for a manual inspection.

An emergency stop is to be provided for the conveyor only. If a board is going through an etch, clean, or rinse cycle, it is to finish those processes, but the conveyor should not start up if the emergency stop is pushed. The stop can be cleared with the start-conveyor push button. Use the following when designing your control.

PB1	Start conveyor	I:1/0
PB2	Stop conveyor only	I:1/1
LS1	Limit switch for detecting when the board is directly over the etching tank	I:1/2
LS2	Limit switch for detecting when the board is directly over the rinse tank	I:1/3
LS3	Limit switch for detecting when the board is directly over the cleaner tank	I:1/4
LS4	Limit switch for detecting the low limit on the lower mechanism	I:1/5

LS5	Limit switch for detecting the high limit on the raise mechanism	I:1/6
LS6	Limit switch for detecting when the board is at the final manual inspection station	I:1/7
	Start-state output	O:2/1
	Conveyor-on-state output	O:2/2
	Conveyor-off-state output	O:2/3
	Lower-state output	O:2/4
	Etch-state output	O:2/5
	Clean-state output	O:2/6
	Rinse-state output	O:2/7
	Raise-state output	O:2/8
	Inspect-state output	O:2/9
	Emergency-stop-state output	O:2/10
M1	conveyor motor	O:2/11
M2	raise motor	O:2/12
M3	lower motor	O:2/13
	Stop-off latch	O:2/14
	Etch timer	T4:1
	Clean timer	T4:2
	Rinse timer	T4:3

Using this description, you can create a state diagram. First, break the process up into small steps that must be accomplished. Show these steps as ellipses with a verbal description of what happens in every step. Each of these is called a state. Secondly, show the Boolean equation that makes a transition between each state. The Boolean statement indicates what control events have to happen to go from one state to another. Figure 10.1 shows the state diagram implementing the process as described. You may have to make several sketches on paper to make the diagram look orderly and to eliminate crossovers.

Once the state diagram is made, you can implement it using several methods. This chapter explores three different techniques: zone-control, subroutines, and sequential function charts. The zone-control and subroutines examples use the SLC 500 and the sequential function chart uses the PLC 500.

▶ 10.2 USING ZONE CONTROL FOR STRUCTURED PROGRAMMING

First, consider zone control. Once you have made the state diagram, the second step is to write the control for all the states. Each state has its own control

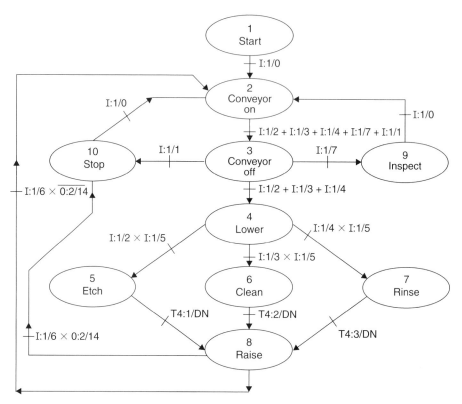

Figure 10.1 State diagram.

bracketed by the zone-control instruction, which is Master Control Reset (MCR). You use a pair of MCRs to bracket the control, one at the beginning and one at the end. The first rung with the MCR output instruction is called the start fence, and it determines if you execute the logic between it and the next MCR, which is called the end fence. If this first MCR sees a true condition, the processor scans and executes the logic between this first rung, the start fence, and the next MCR. If this first MCR sees a false condition, the processor does not scan this control but instead says all outputs are false between the MCRs. This causes nonretentive outputs either to go false or to reset and all retentive outputs to remain in their last state. You can add more structure to the creation on the state ladder logic by allowing every state to have the same general form shown in Figure 10.2.

The first rung, referred to as the start fence, always has the condition for entering the state. Assign an output address that will be active if you are in that state.

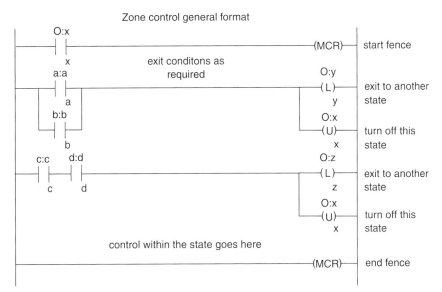

Figure 10.2 General form of implementing a state using the SLC 500.

Turn these outputs off and on to turn each state off and on. The second set of rungs are always the exit conditions for going to a different state; they turn on an output that is associated with a different state. Also, if you ever get an exit condition that causes another state to be active, you will turn off the output for the state you are now executing. If the exit conditions are not met, you will not turn on a new state; if they are, you will. The third set of rungs is used to do whatever needs to be done in the state. These tasks may require only one rung or many rungs, depending on their complexity. Fourth and finally, you will provide the end fence by having an unconditional MCR.

Using this general format ensures that only one state is active at a time. Serious problems can result in sequential control if two states are active at once. For instance, milling a head for an engine and receiving a fixture at the wrong time could cause damage to the part and, potentially, to the crew working with the process.

To implement the control for state 2, use the state diagram (Figure 10.1) and assigned inputs and outputs. First, write the logic for the start fence for state 2.

Second, write the logic for exiting from this state and turning off the present state. Note that + means ORing and × means ANDing.

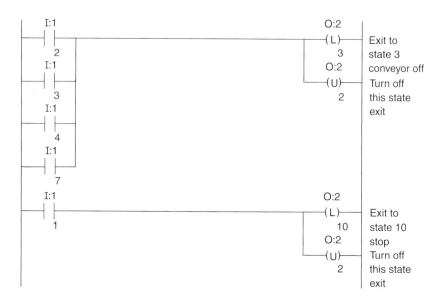

Third, turn the conveyor on when you enter this state.

Finally, create an end fence.

Figure 10.3 shows what a complete conveyor-on zone control should look like.

The raise state is implemented in Figure 10.4, which shows how the emergency stop uses a latch that picks up any time the emergency stop button is pushed. This latch stops the conveyor after the raise motor is finished raising the board so the board will not be stuck in the etch, clean, or rinse tanks. Observe that the general structure is duplicated here: first, the start fence; second, the exit conditions; third, the task you want to accomplish in the state; and fourth, the exit procedure that turns off the state.

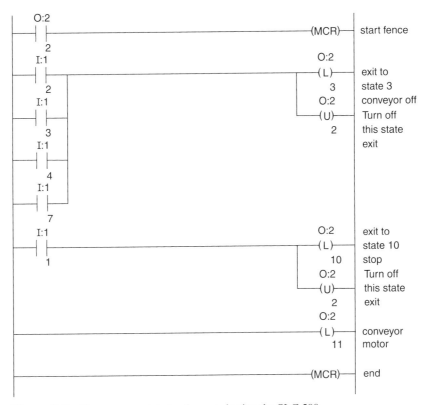

Figure 10.3 Conveyor-on state implemented using the SLC 500.

To get into start state, you use the control program in Figure 10.5. The S1/15 is the first scan bit, which is true only for the first scan. Otherwise, all you do in the start state is to wait for the start push button to be pushed.

▶ 10.3 USING SUBROUTINES FOR STRUCTURED PROGRAMMING

Next, you will implement Figure 10.1 by using subroutines for each state rather than zone control. First, assign outputs to each state. Then, in the main program, simply call a subroutine for every state in the state diagram. To get into the first state at startup, use the first rung in the main program to set an output to get into one of the states. Output O:2/1 is set in the main program in Figure 10.6.

Next, you must make a subroutine program for every state, and these subroutines have to follow the same format. First, put in the rungs for the exit conditions, which will turn on the next state. Determine these exit conditions by looking at the

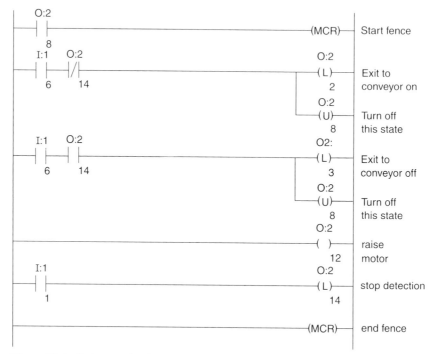

Figure 10.4 Raise state implemented using the SLC 500.

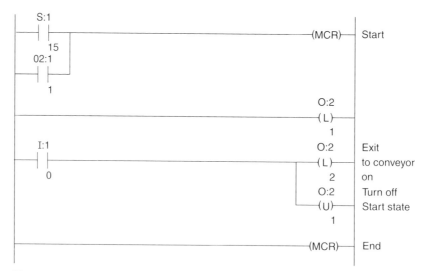

Figure 10.5 Control program for the start state.

154

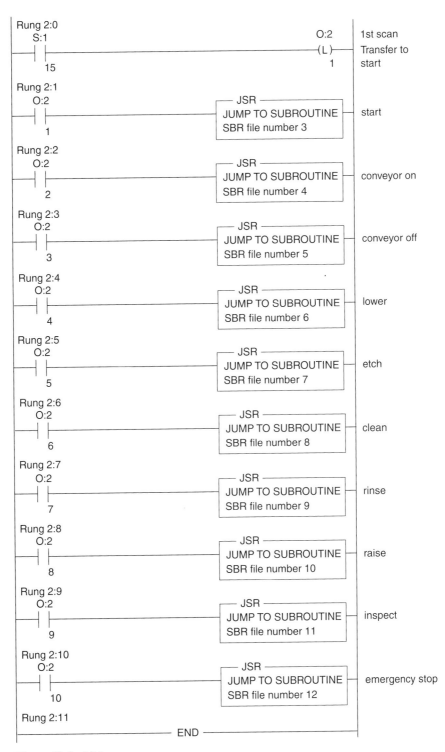

Figure 10.6 Main program.

155

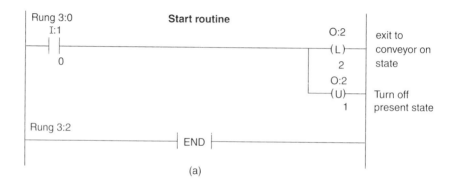

(a)

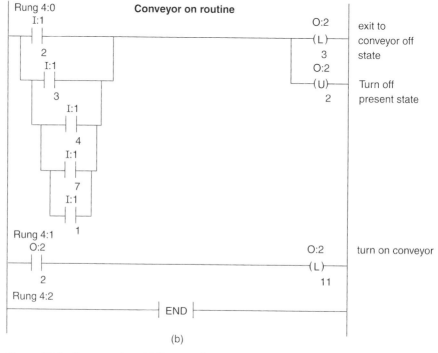

(b)

Figure 10.7   Programming; (a) Start routine; (b) Conveyor-on routine; (c) Etch routine.

state diagram and implementing the required logic. This is the same process you did for zone control. Also, if you exit, you will turn off the present state. Second, do what needs to be done in the state, such as: turning on a timer or starting a motor. Finally, return to the main program if you are exiting. The programming for the start, conveyor-on, and etch subroutines are shown in Figure 10.7.

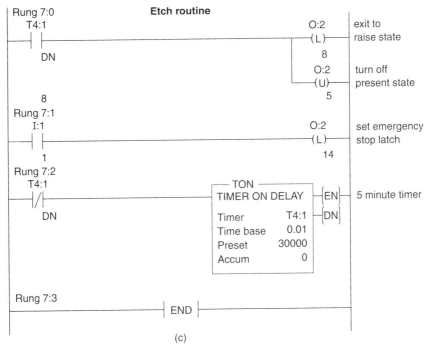

Figure 10.7 Continued.

▶ 10.4 SEQUENTIAL FUNCTION CHARTS

It is important to have a way to handle processes that require more than one state to be active at once as well as a technique for holding the move to the next state until several processes have been completed. One state may be to test a completed assembly such as an engine, but you cannot test the engine until it is completely assembled. There may be hundreds of steps in building an engine; for efficiency, some of these steps can be done simultaneously, but for testing, every simultaneous step must be completed. A graphic picture that shows both sequential and simultaneous processes is called a Petri network.

Allen-Bradley's PLC-5 can implement state or Petri networks using a feature called sequential function charts (SFC). When SFC is used, its charts determine the order of execution of the program the user has written. To use SFC, you must break your program into many smaller programs and then link them together through the SFC. Figure 10.8 shows an example of what a typical chart would look like. Note that if a state is duplicated, it can share the same file number with another state in the SFC.

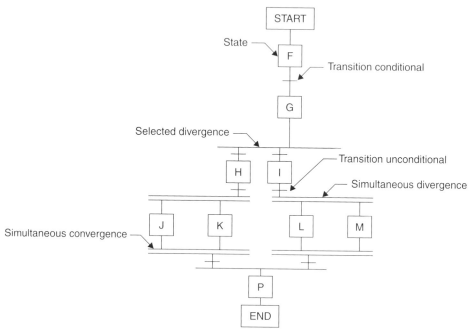

Figure 10.8   A typical sequential function chart.

The following gives definitions for the terms used in Figure 10.8.

State.   This symbol, when active, defines what is to be done through a user-written PLC ladder-logic program. In Figure 10.8, when the processor is first put in the run mode, it will automatically go to the first state, which is F. It will stay there and execute its program until the transition below it is true. Remember, if a transition is duplicated, it can share the same file number with another state in the SFC.

Transition Conditional.   This symbol, when active, defines the conditions for exiting the previous state through a user-written PLC ladder-logic program. The program written for this must have a conditional end-of-transition (EOT) output instruction.

Transition Unconditional.   This symbol causes the previous state to be active for only one scan. Its program simply contains a rung with an unconditional EOT output instruction.

Selected Divergence.   This single line, followed by conditional transitions, lets the user select the *branch* of the SFC that the process will follow. The first transition to be true determines the branch to be followed. If all the transitions are simultaneously true, the process will choose the leftmost branch.

Simultaneous Divergence. Simultaneous divergence works in conjunction with the simultaneous convergence symbol. This double line indicates parallel operation of all branches connected to it. All the branches are entered simultaneously, and the execution of each branch is started.

Simultaneous Convergence. Simultaneous convergence works in conjunction with the simultaneous divergence symbol and requires all branches connected to it to be finished before passing to the next part of the SFC. (*Note:* Simultaneous divergence and simultaneous convergence must be paired together or an error will result.)

10.4.1 Example: Sample Program

A factory makes ceramics. In order for the ceramics to be finished, they must be baked in an oven. The fuel for the oven is oil. The oven is turned on by an external 24-h timer that starts the oven early in the morning.

There is a flame detector that looks for a flame 2 s after the oven is turned on. If the flame detector does not sense a flame and 2 s have elapsed, the control must do two things. First, it must set a lockout latch and second, it must turn off the oven. The system cannot be restarted until this latch is reset manually.

It is unsafe to restart the system until the unburned fuel vapors are purged from the oven; this requires the fire chamber to be purged with fresh air for 30 min. If a flame failure occurs, the system should set off an alarm so maintenance personnel will look at the problem and fix it as quickly as possible. An alarm control will sound an audible alarm and turn on a light. The audible alarm can be shut off by pushing an acknowledge push button. The latch must be reset manually after the system has been examined and fixed. Resetting the latch turns off the alarm light. Another attempt at ignition cannot be made until the fire chamber has been purged because the vapors from the unspent fuel could cause a dangerous explosion; therefore, a restart should not occur until after the 30-min wait to purge. The purge should start immediately after a flame failure. If the purge is completed before maintenance people find the problem, when the problem is resolved the operation should restart immediately after the latch is reset. This will minimize the down time.

Reducing the down time requires two processes to go on at the same time: purging and fixing the problem. These processes are not synchronized, which is sometimes referred to as a parallel asynchronous mode of operation. You can use controls to make sure an unsafe condition does not occur. The states and transitions to use are as follows:

States

Oven on and start 2-s timer
Alarm: light and bell
Alarm: light only
Reset 2-s timer
Set latch
Rest latch
Purge fan
Do nothing
Oven off
Global reset

Transition Devices	Inputs	Outputs
Acknowledge PB	I:000/0	
Flame	I:000/1	
Shutdown latch		O:001/0
Reset shutdown latch	I:000/3	
Flame-out timer	T4:0	
Purge timer	T4:1	
Automatic external timer	I:000/2	
Alarm light		O:001/13

The first step is to go from the verbal description of the problem to a chart that implements the process desired. You will probably need to make several trial sketches with paper, pencil, and eraser. Once you have a sketch you are satisfied with, you can enter the chart using the software. Figure 10.9 shows the SFC for this process. Other charts could be devised that would look different but would still correctly implement the process. The chart in Figure 10.9 is stored in an assigned program file 1 when you use the software to make an SFC.

The second step is to write a program for every step and transition. Each transition and step must be assigned a file number. If a transition or state is duplicated, it can be assigned the same file number. All transition programs must have an EOT. A typical transition program is shown in Figure 10.10 for file 2, titled off-control external timer. A typical state is shown for file 4, oven-on and 2-s timer. Not all the programming is given in the figure, because parts are left for the student to do in the chapter exercises.

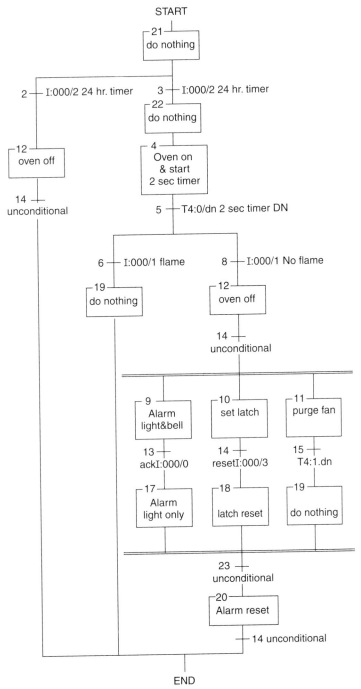

Figure 10.9 SFC for implementing oven control.

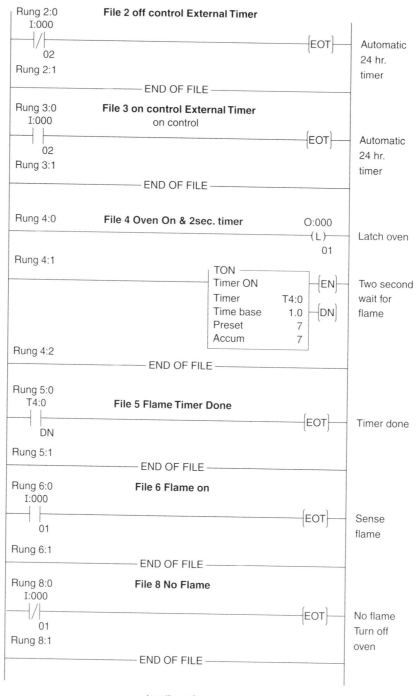

(continues)

Figure 10.10 Typical transition program.

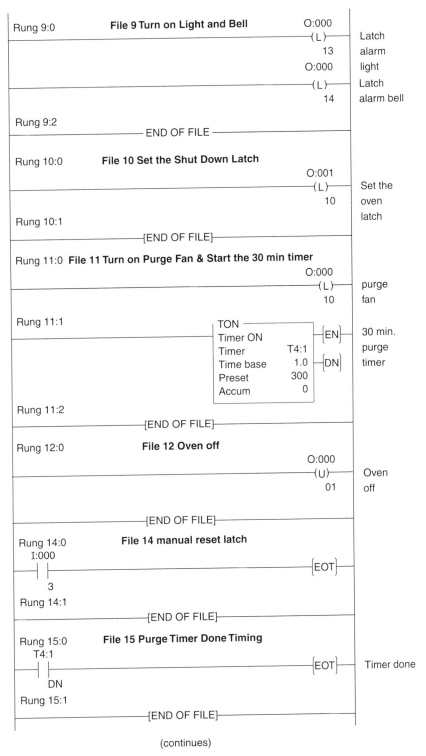

Figure 10.10   Continued.

163

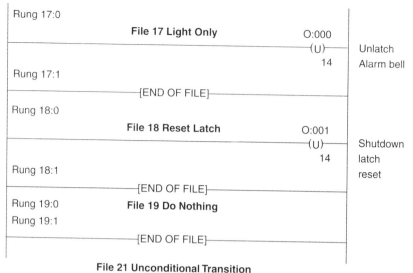

Rung 17:0

File 17 Light Only   O:000
                                                          —(U)—   Unlatch
                                                           14   Alarm bell
Rung 17:1

————————————[END OF FILE]————————————

Rung 18:0

File 18 Reset Latch   O:001
                                                          —(U)—   Shutdown
                                                           14   latch
Rung 18:1   reset

————————————[END OF FILE]————————————

Rung 19:0   **File 19 Do Nothing**
Rung 19:1

————————————[END OF FILE]————————————

File 21 Unconditional Transition

Figure 10.10   Continued.

10.4.2 SFC Scanning

Finally, consider how the SFC is scanned. Examine Figure 10.11, which you can
use as a guide.

Figure 10.12 gives the order of the scanning. First, S1 is scanned, followed by
transition T1. If T1 is true, then S1 is postscanned (pscan). In the default setup, the
S1 postscan forces all rungs in the state to false and adjusts outputs accordingly.
This means *nonretentive outputs* turn off or reset and *retentive outputs* stay in their
last state. If T1 is false, the processor checks the next transition in this selected divergence instruction, which is T2. If T2 is true, then S1 is postscanned (pscan). If
it is false, it checks the next transition in this selected divergence instruction, which
is T3. If T3 is true, then S1 is postscanned (pscan). If it is false, it goes back to scan
S1 and repeats the process. You should note that it starts with the leftmost transition and scans from left to right for selected divergence.

Following Figure 10.10, if T1 is true, work through the leftmost branch per
Figure 10.13.

Following Figure 10.12, if T3 is true, work through the center branch per Figure 10.14. Notice the center branch contains a simultaneous divergence, and you
enter this parallel operation when T6 is true. Once in the simultaneous divergence,
states S5, S6, and S7 are scanned in succession, starting with the leftmost state and
ending with the rightmost state. You have to complete all the branches between the

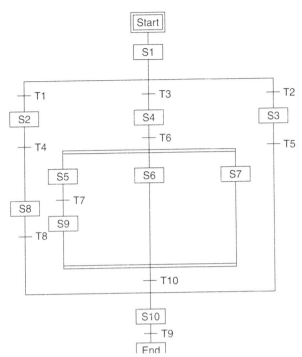

Figure 10.11 SFC scanning diagram.

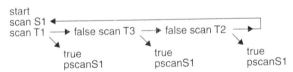

Figure 10.12 Order of SFC scanning.

simultaneous divergence and simultaneous convergence before you can go on. This will happen until T7 goes true; then S6, S7, and S9 are postscanned. Finally, the process goes to state S10 and starts checking T9.

Following Figure 10.12, if T2 is true, work through the rightmost branch per Figure 10.15.

Sequential function charts have the same advantages as state and Petri networks because SFC is a structured approach. Once you put the time and effort into programming this way, the benefits will follow.

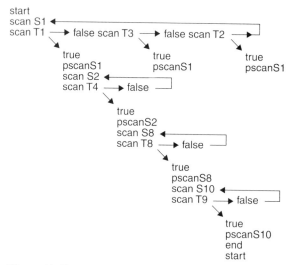

Figure 10.13 Scan if T1 is true.

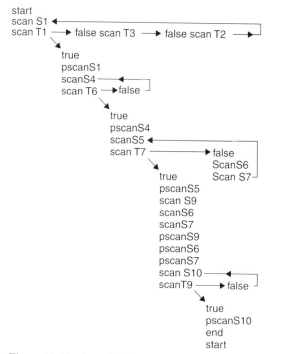

Figure 10.14 Scan if T3 is true.

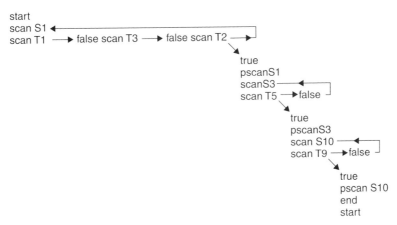

Figure 10.15 Scan if T2 is true.

SUMMARY

It takes time and effort to learn structured methods of programming, but it has many benefits. Advantages:

1. It automatically creates good documentation.
2. It breaks up control into smaller, more manageable parts.
3. Troubleshooting is easier.
4. The process is easy to watch via a PC monitor.
5. Efficiency is increased, thus shortening the time needed to solve complex problems.
6. Scan time is reduced because only the active states are scanned.

► EXERCISES

1. Given the state diagram in Figure 10.1, write a program for the lower state using zone control.
2. Given the state diagram in Figure 10.1, write a program for the etch state using zone control.
3. Given the state diagram in Figure 10.1, write a program for the clean state using zone control.
4. Given the state diagram in Figure 10.1, write a program for the rinse state using zone control.

5. Given the state diagram in Figure 10.1 and the main program, Figure 10.6, write a program for the conveyor-off subroutine.

6. Given the state diagram in Figure 10.1 and the main program, Figure 10.6, write a program for the lower subroutine.

7. Given the state diagram in Figure 10.1 and the main program, Figure 10.6, write a program for the clean subroutine.

8. Given the state diagram in Figure 10.1 and the main program, Figure 10.6, write a program for the rinse subroutine.

9. Given the state diagram in Figure 10.1 and the main program, Figure 10.6, write a program for the raise subroutine.

10. Given the state diagram in Figure 10.1 and the main program, Figure 10.6, write a program for the inspect subroutine.

11. Given the state diagram in Figure 10.1 and the main program, Figure 10.6, write a program for the emergency-stop subroutine.

12. When you leave a state in SFC, what happens?

13. When using SFC, how does the scan go through the chart?

14. Given the sequential function chart in Figure 10.9, write a program for:

   File #20 Alarm Reset

15. Given the sequential function chart in Figure 10.9, write a program for:

   File #13 Acknowledge Push Button

16. Given the control problem described next, do the following:
   a. Draw a state diagram.
   b. Show the main program for implementing this control using subroutines instead of zone control.
   c. Show the ladder logic for the conveyor-on state only.
   d. Document all rungs, input devices, and output devices.

A chassis is placed on a conveyor and a painting process is initiated by SLC 501 when an operator pushes the start push button. The chassis is manually placed on a conveyor and travels to the first dip tank via the conveyor. When it is directly over the dip tank, it is lowered into the cleaning tank by energizing a lowering motor. The lowering motor drives a cable that operates a mechanism to lower it a set distance into the tank. A limit switch on this mechanism provides a signal to shut off the motor. It remains in the cleaning solution for 30 s and then is removed. The raise mechanism has a limit switch for signaling when it has been raised to the proper height so that it can move via the conveyor to the primer process. It is then dipped for 2 min. It must be lowered and removed from the tank in the same manner as with the cleaning process. Finally, it is put in a paint tank for 240 s, and then it goes to an inspection station via the conveyor for a manual inspection. The air drying after each dip is accomplished by waiting an appropriate amount of time after the chassis is raised out of the tank, so the conveyor should delay starting to allow time for the part to dry. Use the following I/O and timers when designing your control. If a chassis is going through a clean, primer, or paint cycle, it is to finish those processes, but the conveyor should not start up if the emergency stop is pushed. The emergency stop can be cleared with the start-conveyor push button.

Steps	Time	Timers	Limit Switch	Address
0 Start			PB1	I:1/0
1 Dip (cleaner)	30 s	T4:0	LS-1	I:1/1 at clean tank
2 Air dry	120 s	T4:1		
3 Dip (primer)	120 s	T4:2	LS-3	I:1/3 at primer tank
4 Air dry	60 min	T4:3		
5 Dip (paint)	240 s	T4:4	LS-5	I:1/5 at paint tank
6 Air dry	120 min	T4:5		
7 Inspect station			LS-7	I:1/7
8 Lower limit			LS-8	I:1/8
9 Raise limit			LS-9	I:1/9
10 Emergency stop				I:1/10
Start-state output				O:2/1
Conveyor-on-state output				O:2/2
Conveyor-off-state output				O:2/3
Lower-state output				O:2/4
Clean-state output				O:2/5
Primer-state output				O:2/6
Paint-state output				O:2/7
Raise-state output				O:2/8
Dry				O:2/9
Inspect-state output				O:2/10
M1 conveyor motor				O:2/11
M2 raise motor				O:2/12
M3 lower motor				O:2/13
Emergency-stop latch				O:2/14
Emergency-stop-state output				O:2/15

17. Using sequential function charts, design a push-button lock that will operate a solenoid to unlatch a door for 20 s after three separate push buttons have been

pushed in the correct sequence. If the wrong sequence is pushed at any time, the program will prevent a further try for 10 s and then permit another try. If more than two consecutive tries result in errors, then the lock is to be disabled or locked out. A push button is to be provided to reset the lock if it is locked out. Use the following I/O and file numbers:

States	File Number	Transition From	File Number
1st try	2	1st try	7, 8
2nd try	3	2nd try	9,10
3rd try	4	3rd try	11,12
Error	5	Error	13
Operate	6	Operate	14

Transition Device	Inputs	Outputs
PB-1 (first correct)	I:000/0	
PB-2 (second correct)	I:000/1	
PB-3 (third correct)	I:000/2	
PB-4 (reset)	I:000/3	
Timer 1		T4:1
Timer 2		T4:2
Counter		C5:1

18. Using sequential function charts, design the control for an annunciator panel with a four-alarm system shown in Figure 10.16.

The two processors are located in two different buildings, and each has a local annunciator panel near the processor similar to the one in Figure 10.16. The processors are to communicate over a DH+ network. Each processor needs to see what is happening to the remote processor alarm inputs. You do not want the remote processor to change the input table of the local processor. Therefore, each processor should have a rung that copies word 0 of its input table into B3:0. If a remote processor needs to know what is happening at the other processor's input, it will read word B3:0. Use the message instruction to communicate with the remote processor.

If any alarm input goes high, the following should happen. First, an audible alarm sounds and a flashing light comes on behind a translucent window. The translucent window contains printing to indicate which alarm is active.

Second, the person responding to the system first turns off the audible alarm by pressing an acknowledge push button. This turns off the audible alarm, but the light still remains flashing behind the appropriate window. Once the audible alarm is on, only the acknowledge button can turn it off. Each input can activate the audible alarm and must be acknowledged separately.

Third, if a person fixes the problem or the problem ceases to exist, the light behind the appropriate window goes from flashing-on to continuous-on.

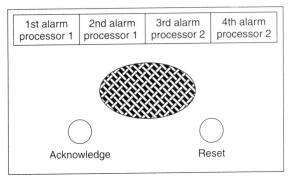

Figure 10.16 Exercise 18.

Fourth, the only way to turn off the lights is to press a reset button. The reset button should momentarily interrupt the power to the alarm system. If everything is fixed, the alarm lights and the audible alarm go off and stay off. If problems still exist, the audible alarm goes off again and lights flash.

EET377 Alarm-Panel Bit Assignment

Processor	1	1	2	2
Alarm-bit assignment	I:000/00	I:000/01	I:000/00	I:000/01
After copy to B3:0	B3:0/00	B3:0/01	B3:0/00	B3:0/01
Read source for other processors	B3:0	B3:0	B3:0	B3:0
Acknowledge PB bit assignment	I:000/2	I:000/2	I:000/2	I:000/2
After copy to B3:0	B3:0/2	B3:0/2	B3:0/2	B3:0/2
Reset PB bit assignment	I:000/3	I:000/3	I:000/3	I:000/3
After copy to B3:	B3:0/3	B3:0/3	B3:0/3	B3:0/3

Bit Assignments

O:000/00	al#1		I:000/00	al#1 (local)
O:000/01	al#2		I:000/01	al#2 (local)
O:000/02	al#3		I:000/02	Acknowledge PB (local)
O:000/03	al#4		I:000/03	Reset PB (local)
O:000/04	al#1 alarm light		I:000/04	Network ready PB(local)
O:000/05	al#2 alarm light			

O:000/06	al#3 alarm light	B3:1/00	al#3 (remote)
O:000/07	al#4 alarm light	B3:1/01	al#4 (remote)
O:000/10	Horn output	B3:1/02	Acknowledge PB (remote)
		B3:1/03	Reset PB (remote)
		B3:1/04	Network ready PB (remote)
B3:2/00	al#1 memory latch	B3:2/09	al#1 acknowledge unlatch once
B3:2/01	al#2 memory latch	B3:2/10	al#1 acknowledge unlatch once
B3:2/02	al#3 memory latch	B3:2/11	al#1 acknowledge unlatch once
B3:2/03	al#4 memory latch	B3:2/12	al#1 acknowledge unlatch once
B3:2/04	al#1 acknowledge latch		
B3:2/05	al#2 acknowledge latch		
B3:2/06	al#3 acknowledge latch		
B3:2/07	al#4 acknowledge latch		
B3:2/08	Flash bit		

19. A water tank has four sensors to detect four different water levels: F, full; M, medium; L, low; and E, empty. There are three pumps, P1, P2, and P3, to fill the tank. When the automatic control is turned on, the following should happen:

a. An empty tank should result in all pumps being turned on until the tank is full.

b. A full tank or water level between full and medium should keep all pumps turned off.

c. A condition below medium but not low should turn on one pump until full is reached.

d. A condition below low but not empty should turn on two pumps until full is reached.

e. If the water level goes below empty while both pumps are on, all three pumps should be on until full is reached.

f. To reduce wear during these fill cycles, the pumps are to alternate through three sequences according to the following table. The second-on pump and the third-on pump should come on if the level sensors call for more pumps. If the first pump comes on and fills the tank without needing the other pumps, the control should switch to the next sequence the next time a pump is required.

	First On	Second On	Third On
Sequence 1	Pump 1	Pump 2	Pump 3
Sequence 2	Pump 2	Pump 3	Pump 1
Sequence 3	Pump 3	Pump 1	Pump 2

Sensors produce a contact closure when they touch the water.

Use a counter to determine when to alternate the pumps. The sensors are connected to I:1/0, I:1/1, I:1/2, and I:1/, and the following list shows the condition of the sensors for the four possible water levels. If there is no bar above the letter, that sensor will send a high to the input.

$\bar{E}$	$\bar{L}$	$\bar{M}$	$\bar{F}$	Completely empty
E	$\bar{L}$	$\bar{M}$	$\bar{F}$	Between empty and low
E	L	$\bar{M}$	$\bar{F}$	Between low and medium
E	L	M	$\bar{F}$	Between medium and full
E	L	M	F	Completely full

Make a state diagram and then implement the state diagram using subroutines.

PART 2

Advanced Programmable Controllers

Communication with Other Programmable Controllers and Computers

CHAPTER OUTLINE

11.1   DH-485™ Communications
11.2   RS-232 and SCADA Communications
11.3   Data Highway Plus™
11.4   ControlNet™
11.5   Ethernet®

LEARNING OBJECTIVES

Upon reading this chapter students should be able to:

- Explain communication concepts.
- Explain DH-485 communications.
- Explain RS232 communications.
- Explain Data Highway Plus communications.
- Explain ControlNet communications.
- Explain Ethernet communications.

INTRODUCTION

Various methods are used to communicate between programmable controllers and between programmable controllers and computers.

This chapter discusses a number of different methods of communication, including **DH-485**™, *RS232*, **Data Highway Plus**™ (DH+™), **ControlNet**™, and **Ethernet**®.

The communication networks are used for supervisory control, data gathering, monitoring device and process parameters, and download and upload of programs.

► 11.1 DH-485 COMMUNICATIONS

DH-485 is an Allen-Bradley proprietary network. It is part of the SLC fixed, 5/01, 5/02, and 5/03 controllers. With an advanced interface converter, the SLC 5/03, 5/04, and 5/05 can communicate through their serial port to DH-485. DH-485 is also used with other devices, such as human-machine interfaces and data table access modules.

DH-485 supports the interconnection of a maximum of 32 devices with a maximum network length of 4000 ft. It has multimaster capability using token-passing access control.

Two types of devices are supported on the network, *initiators* and *responders.* All initiator-type devices can initiate message transfers. Responders receive messages. Some devices, such as the 5/03 or 5/04, are capable of being initiators and responders. A personal computer running programming software is strictly an initiator, whereas a fixed controller is just a responder.

DH-485 is a token-passing network. The initiator that has the token has a right to transmit to a responder. The node holding the token can send a valid packet onto the network. Each time a node has a token, it can send one transmission. The node is allowed two retries. After the node has sent a packet, it tries to send a *token pass* to its successor by passing the token to the next-higher node number. If the node fails to pass the token after three attempts, it will try to find a new successor. There can be a maximum of 32 (0–31) nodes on the network and a minimum of one initiator. The valid node numbers for initiators are 0–31; for responders they are 1–31. You may set the maximum node on the network in the software so that the controller does not look for nodes higher than the maximum. The default maximum node is 31. The SLC 500 fixed and 5/01 controllers can be selected by a maximum of two initiators at the same time. More than two initiators can cause communication timeouts.

Figure 11.1 shows a DH-485 network consisting of three nodes, the computer and two SLC 500 controllers. The 1747-PIC box connects to the serial port on the computer and plugs into the link coupler (1747-AIC) module. The fixed, 5/01, 5/02, and 5/03 controllers have a DH-485 port on the front of the controller. The

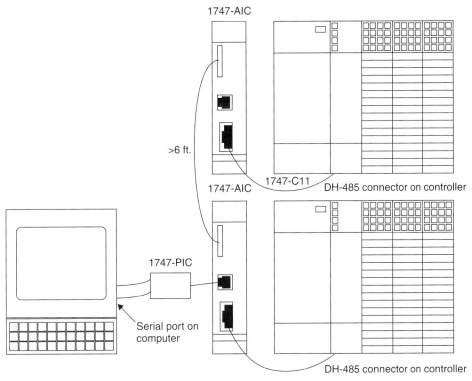

Figure 11.1 DH-485 network.

maximum distance from the first to the last link coupler is 4000 ft. The minimum distance between any two link couplers is 6 ft.

An advanced interface converter (1761-NET-AIC) may be used to connect 5/03, 5/04, and 5/05 controllers to DH-485 through their serial port, as shown in Figure 11.2. When the serial port is used for DH-485 communication, it must be configured for DH-485 using the programming software. The advanced interface converter receives its power, 24 V dc, from the chassis power supply.

▶ 11.2 RS-232 AND SCADA COMMUNICATIONS

RS-232 is an Electronics Industries Association (EIA) standard that specifies the electrical, mechanical, and functional characteristics for serial binary communication. It is included in SCADA (Supervisory Control and Data Acquisition) applications. SCADA applications refer to communications over long distances.

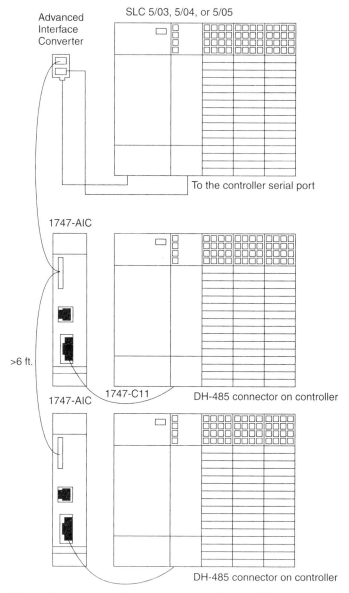

Figure 11.2 Advanced interface converter (1761-NET-AIC).

Using RS-232 from a computer connected directly to a ControlLogix or SLC 500 controller requires no special hardware other than the proper cable. The maximum distance between the computer and controller is 50 ft. Data communication is slower than with the other networks, but RS-232 provides an easy, direct connection between the computer and the controller. Programming changes and monitoring the project for troubleshooting purposes can be done using a serial connection.

Another advantage of serial communications is that with the use of telephone and radio modems, the communication distances between certain system devices are virtually limitless.

The serial port on the SLC 5/03, 5/04, and 5/05 controllers support the following protocols:

- Full-duplex DF1
- Half-duplex DF1 (SCADA)
- DH-485
- ASCII communications

The ControlLogix controller supports the following:

- No handshake
- Full-duplex DF1
- Half-duplex DF1

DF1 protocol allows for communication between devices that support DF1 protocol. Examples of such devices are programming terminals, display terminals, and communication modules.

Full-duplex DF1 protocol is also referred to as DF1 point-to-point protocol. In full-duplex mode there can be simultaneous communication between two devices in two directions. This mode is typically used for communication between the controller and the programming terminal.

Figure 11.3 shows a programming terminal and an SLC 500 communicating from serial port to serial port using DF1 full-duplex protocol. The only additional equipment needed is the 1747-CP3 serial cable for communications. The maximum possible distance between the computer and SLC is 50 ft. The same communication setup can be used between a ControlLogix controller and a programming terminal using a 1756-CP3 serial cable.

Figure 11.4 shows the SLC 500 connected to a computer using modems. Modems used with DF1 full-duplex protocol must be capable of simultaneous bidirectional communication. Dial-up modems designed to be connected to telephone lines are typically full-duplex.

Figure 11.5 shows a master computer connected to slave controllers using half-duplex protocol. DF1 half-duplex protocol uses a master-slave-type communication.

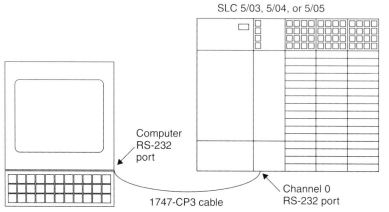

Figure 11.3   RS-232 DF1 full-duplex protocol.

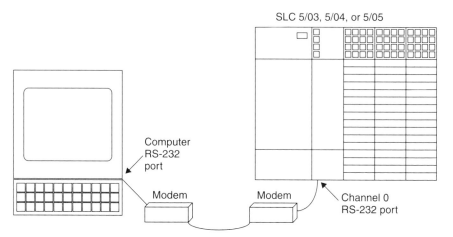

Figure 11.4   Full-duplex communication.

Communication takes place in one direction at a time. There may be one master and up to 254 slave node addresses. When there is only one slave connected to a master, the master may be directly connected to the slave.

   The SLC 5/03 (OS301) and higher, 5/04, and 5/05 controllers allow for ASCII communications through the serial port when the port is configured for user mode. When applying the user mode, all received data are placed in a buffer and the data are accessed using the ASCII instructions in the ladder-logic program. ASCII string data can also be sent to most devices that accept ASCII protocol. Figure 11.6 shows a device that accepts ASCII protocol connected to the serial port of a SLC controller.

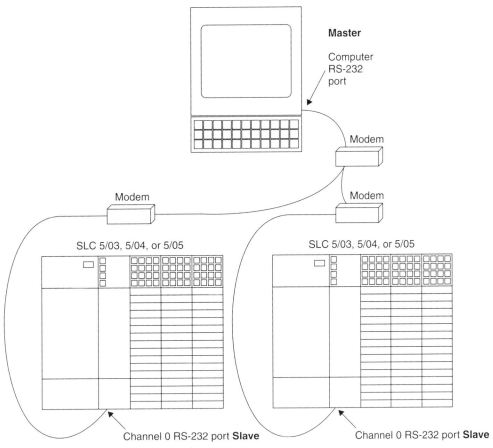

Figure 11.5 DF1 Half-duplex protocol.

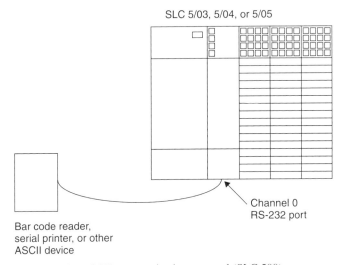

Figure 11.6 ASCII communication protocol (SLC 500).

▶ 11.3 DATA HIGHWAY PLUS (DH+)

Data Highway Plus (DH+) is a local information network that is designed for remote programming and for accessing and transferring data. It is the primary communications network for the SLC 5/04 and PLC-5® controllers. Using a 1756-DHRIO bridge module with a ControlLogix system allows communication with other DH+ devices, but the primary communication methods in a ControlLogix system are Ethernet, ControlNet, and DeviceNet.

DH+ is a token-passing network with a floating master. The device with the token may initiate data transfer. The token is passed sequentially from node to node according to the node's address. A maximum of 64 nodes occur on a local network, and the nodes are numbered from 0 to 77 octal. Three different *baud rates* are available on DH+: 57.6k baud (maximum cable length 10,000 ft), 115.2k baud (maximum cable length 5000 ft), and 230.4k baud (maximum distance 2500 ft).

Figure 11.7 shows the DH+ connectors for an SLC 5/04 controller. The round DIN connector is typically connected to the programming terminal, and the three-pin Phoenix connector is used to form a daisy-chain connection between the SLC 5/04 and other DH+ devices. A communication card is required in the computer to enable the computer to communicate on DH+. The type of card used depends on the computer type. The most common cards are the following:

- 1784-KTX, -KTXD   IBM XT/AT computer bus   DH+ or DH485
- 1784-PCMK   PCMCIA slot in computer   DH+ or DH485
- 1784-KT/B   IBM XT/AT computer bus   DH+

▶ 11.4 CONTROLNET COMMUNICATIONS

ControlNet is the primary communications network for ControlLogix controllers. It combines the capabilities of DH+ and remote I/O networks. ControlNet is an open, high-speed, deterministic network that transfers time-critical I/O updates, controller-to-controller interlocking data, and non-time-critical data (such as data monitoring and program uploads and downloads) on the same network. Time-critical data are referred to as scheduled data, and the non-time-critical data is referred to as unscheduled data.

ControlNet can be used to connect controllers, computers, coordinated drive systems, weld controllers, motion controllers, and human-machine interfaces (HMI). The ControlNet network transfers scheduled data at 5M bits/s. There can be up to 99 addressable nodes on a link, and the ControlNet network can be accessed from every node on the network. Multiple controllers can control their own outputs on a single common ControlNet link, and they can monitor all inputs on the link. Also, any controller on the link can send a message to any other controller on the link.

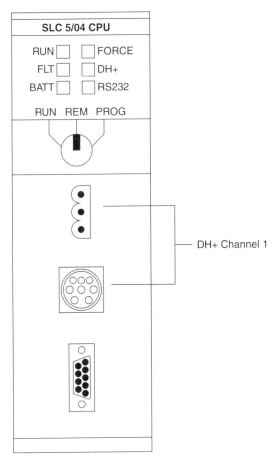

Figure 11.7 DH+ connectors for an SLC 5/04.

Both discrete and nondiscrete transfers of data can be used on a ControlNet link. A discrete transfer of I/O data occurs when 8, 16, or 32 bits are transferred to or received from each I/O module in an I/O chassis. All the I/O modules in a chassis can be updated with one discrete transfer. A nondiscrete transfer of I/O data occurs when a block with a maximum of 64 words of data is transferred to an I/O module.

Figure 11.8 shows a ControlNet cable system. The taps connect any device with the ControlNet cable system. The repeater connects one segment to another. A link is a group of nodes with unique addresses that have a range from 1 to 99. A segment consists of trunk cable sections connected via taps with no repeaters and a terminator comprising a 75-Ω resistor mounted in a BNC plug. A bridge is a device that connects one link to another. The standard cable that can be used to construct a network is a quad-shield RG-6-type coax cable. The maximum length of a

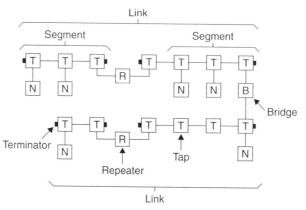

Figure 11.8 ControlNet cable system.

segment depends on the number of taps in the segment and the type of cable used. A repeater allows for increasing the number of taps and extending the length. A network is the collection of nodes that are connected together using repeaters and bridges.

▶ 11.5 ETHERNET

Ethernet is a popular plantwide information network, including both the plant floor and office communications. It is a local-area network that is designed for the high-speed exchange of data between computers and other Ethernet devices. The advantage of using Ethernet is that a wide variety of products supplied by various manufacturers is available to use to communicate over long distances.

The Internet uses Ethernet with a *TCP/IP* protocol (transmission-control protocol/internet protocol). Due to the fact that Ethernet technology and TCP/IP protocol have been made public, software and physical media are readily available. On top of the TCP/IP protocol, the SLC5/05 and the ControlLogix modules utilize the control and information protocol (CIP) to provide real-time I/O messaging and information/peer-to-peer messaging. This combination is referred to as Ethernet industrial protocol (Ethernet/IP).

A block of *IP addresses* is assigned to each manufacturer by the IEEE, and the manufacturer assigns a unique address to each Ethernet device. The IP address uses a 48-bit addressing scheme. Each data packet contains a 48-bit destination and a 48-bit source address. The destination address may be for a single device or a multicast address to be received by multiple Ethernet devices on the network. Using this scheme, a device can ignore traffic that is not intended for them. This greatly increases the speed of the network.

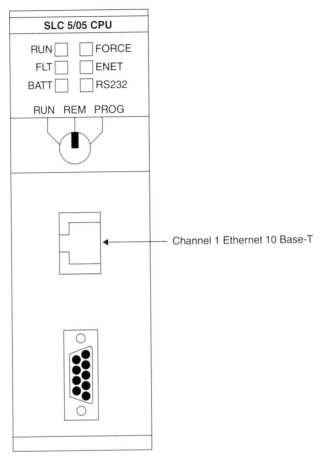

Figure 11.9 SLC 5/05 Ethernet controller.

SLC 5/05 and ControlLogix controllers may use Ethernet communications. Ethernet is built in the SLC5/05 controller and ControlLogix uses a 1756-ENET (10M-bit/s) or a 1756-ENBT (10M/100M-bit/s) bridge module to communicate on Ethernet.

Figure 11.9 illustrates an SLC5/05 controller. The SLC5/05 controller uses channel 1 as an Ethernet communication channel. It is a 10 base-T, RJ45 Ethernet connector.

Figure 11.10 shows an Ethernet network with a ControlLogix 1756-ENET module, personal computer, and SLC5/05 connected to a hub or switch. The multiport hub or switch enables the devices to be connected in a star topology. The hub or switch acts as a concentrator for connecting multiple devices. The hub or switch and its connected devices may make up the Ethernet network or the hub may be linked to another hub or to a fiber-optic backbone that spans the facility.

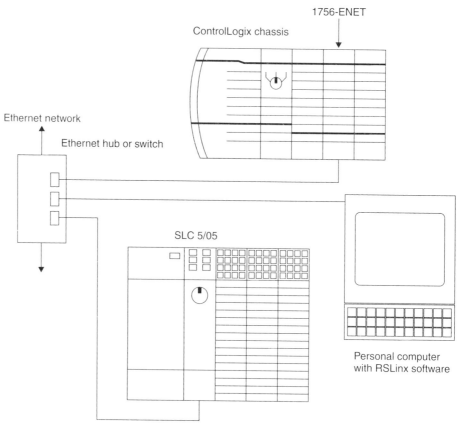

Figure 11.10   Ethernet network.

SUMMARY

The SLC 500 and ControlLogix controllers incorporate various types of communications between the controllers and between controllers and computers.

DH-485 provides the primary communications when using the fixed, 5/01, 5/02, and 5/03 controllers. DH-485 can be used for communication between SLCs or between SLCs and computers.

DH+ supplies the primary communications of the SLC 5/04 controller and can also be used in ControlLogix communications.

Serial communications are used with the SLC 5/03, 5/04, and 5/05 controllers and ControlLogix controllers to communicate with computers or other serial devices.

ControlNet is the primary communication method of the ControlLogix controllers and combines the function of DH+ and remote I/O.

Ethernet is used with the SLC 5/05 controller and also with the ControlLogix controller. It can be used to communicate on a plantwide computer network.

► EXERCISES

1. What is the maximum number of nodes on a DH-485 network?
2. What is the maximum cable length on a DH-485 network?
3. When connecting fixed, SLC 5/01, 5/02, or 5/03 controllers together on a DH-485 network, it is necessary to use what device for communications?
4. What is the advantage of using the serial port to connect a computer to an SLC or ControlLogix controller?
5. What is the disadvantage of using the serial port to connect a computer to an SLC or ControlLogix controller?
6. What ControlLogix module performs DH+ communications?
7. What SLC controller has built-in DH+ communications?
8. What is the maximum number of nodes on a DH+ network?
9. What are the baud rates and maximum communication distances for DH+?
10. What ControlLogix module is used for Ethernet communications?
11. What SLC controller has built-in Ethernet communications?
12. What is the communication rate of scheduled data on an Ethernet network?
13. What is the maximum number of nodes on an Ethernet link?

Automatic Control Using PID Instruction

CHAPTER OUTLINE

LEARNING OBJECTIVES

Upon reading this chapter students should be able to:

- Determine the information needed to scale analog I/O or numeric data being transferred between devices and instructions.
- Describe the three types of control programs used for PID execution and explain how they differ.
- Correctly enter data required to use a PID instruction.
- Determine PID tuning using the direct synthesis–step response method.
- Determine PID tuning using the Ziegler-Nichols open-loop test.
- Determine PID tuning using the Ziegler-Nichols closed-loop test.

INTRODUCTION

You have learned that PLCs can do sequential control; this chapter considers automatic control. Automatic control requires feedback from what you are trying to control. The control of speed in your car is a good example. A computer reads a sensor that detects speed, it checks the speed against a desired set point, and then it sends a correction by increasing or decreasing the gas fed to the motor. It continues to check the sensor that detects the speed to see if it is closing on the **set point.** If it is not, it sends a new correction to the gas feed. It constantly keeps checking and adjusting; this process is called **automatic control.**

▶ 12.1 SCALING

You need to learn about the use of analog input and analog output modules before you use the PID instruction. Consider an SLC 500 analog I/O. The *transducers* are used to convert various types of analog information to electrical signals. The most popular electrical signals are voltages in the range of ± 5 V or ± 10 V and currents in the 4- to 20-mA range. The following are some examples of the many different types:

Pressure

Temperature

Distance

Light intensity

Speed

Frequency

Distance

There are many other types of possible transducers, which are devices that produce either voltage or current proportional to some engineering units such as temperature (°C or °F), pressure (lb/in.^{2}), distance (cm), etc.

When you use analog I/O, you will need to scale in order to make it compatible with other devices or to put raw numeric data into engineering units. *Scaling* is based on the algebraic expression for a linear relationship, which is given as the equation $y = mx + b$. When scaling, x is what you want to change and y is what you want to change to. It helps to make a graph such as Figure 12.1.

You scale for the following reasons:

1. To change the way numerical information is presented. For example, you want to change raw machine data, which is not very meaningful to humans, to engineering units that are easy to understand. Some refer to scaling as going from raw,

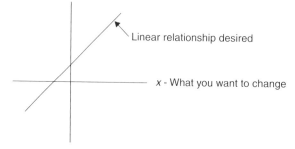

Figure 12.1 Graph of a linear relationship.

unitless data to engineering units and descaling as going from engineering to raw, unitless data, but it is really not a different process mathematically. It is just a matter of which quantity is represented by x and y.

2. So that the two devices can interchange numerical information. Scaling ensures that the range of each device is not exceeded and as much resolution as possible is maintained. These exchanges are from unitless to unitless data or unit-based to unit-based data.

Figure 12.2(a) illustrates changing engineering units to unitless. You need to calculate the slope, or rate of change, and the offset to be able to write an equation for the line. The range of y is 0 to 16,384 and the range of x is 32 to 212.

$$m = \frac{y_{max} - y_{min}}{x_{max} - x_{min}} = \frac{16{,}384 - 0}{(212 - 32)°F} = 91.02/°F$$

Figure 12.2(a) shows when x is 212, then y is 16,384.

$$\text{offset} = b = y - mx = 16{,}384 - 91.02/°F \times 212°F = -2912.24$$
$$y = mx + b = (91.02/°F)x - 2912.24$$

Figure 12.2(b) shows changing unitless to engineering units. You must calculate the slope or rate of change and the offset to be able to write the equation for the line.

$$m = \frac{y_{max} - y_{min}}{x_{max} - x_{min}} = \frac{(212 - 32)°F}{16{,}384 - 0} = 0.010986°F$$

Figure 12.2(b) shows that when x is 16,384, then y is 212.

$$\text{offset} = b = y - mx = 212°F - 0.010986°F \times 16{,}384 = 32°F$$
$$y = mx + b = (0.010986°F)x + 32°F$$

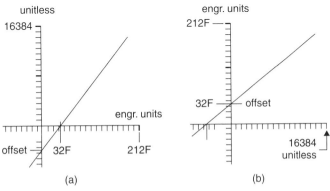

(a)   (b)

Figure 12.2 Changing units; (a) When x is 212, then y is 16,384; (b) When x is 16,384, then y is 212.

To change 150°F to a unitless measure, use the equation $y = mx + b = (91.02/°F)x - 2912.24$. Then,

$$y = mx + b = 91.022/°F \times 150°F - 2912.24 = 10,740$$

To change 10,740 back to °F, use the equation $y = mx + b = (0.010986°F)x + 32°F$:

$$y = mx + b = (0.010986°F)x + 32°F = 0.010986°F \times 10,741$$
$$+ 32°F = 150°F$$

Figure 12.3 shows a typical application, in which a transducer is measuring the temperature of liquid in a tank and feeding this information into an SLC 500 analog input module. The job of the analog input module is to convert this information into binary so that an SLC 500 processor can use it. There are three types of units in this diagram:

Engineering units: The units a human uses and understands
Transducer units: Either a voltage or current
Binary or machine units: The units the processor needs

The line drawn in Figure 12.3 represents the relationship between engineering units and binary or machine units. Recall that the algebraic equation for a line is

$$y = mx + b$$

Substituting the units for x and y gives the following equation:

$$°C = slope \times binary + offset$$
$$y_{max} - y_{min} = 200°C - (-100°C) = 300°C$$
$$x_{max} - x_{min} = 32,767 - (-32,768) = 65,535$$
$$offset = 50°C \quad when \ x = 32,767, y = 200°C$$
$$°C = 0.004577°C \times binary + 50°C$$

Figure 12.4 shows how to solve these equations. The sample program is for an SLC 500 series processor. The source is from an analog input module in slot 1, and you are reading channel 1. You can use an SLC 500 NIO4V analog I/O module, which has a full-scale integer range of $-32,768$ to $32,767$, to read the input. The analog input module provides the processor with a 16-bit binary number via I:1.1 and after the program is executed N7:0 displays the integer value for engineering units, which in this example is temperature in degrees Celsius.

This set of instructions scales the channel 1 analog input value from an analog input module in slot 1.

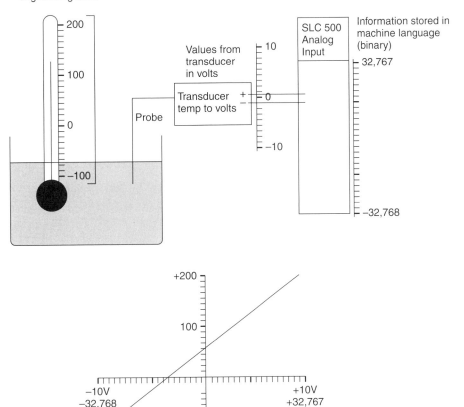

Figure 12.3 Measuring temperature.

Figure 12.5 does the same thing as Figure 12.4, but it is condensed into one instruction. It is available on the SLC 502 and above. It is called the Scaling (SCL) instruction. You need to know the slope, or rate, and the offset. The rate is $300/65,535 = 0.004577$, but you need to write this decimal value as an integer so it can be stored in an integer file. This is done by multiplying by 10,000: rate $\times$ $10,000 = 45.77$. Then rounding to 46 so that you have an integer input. The SCL instruction multiplies the source value by the rate value, divides by 10,000, and adds the offset. If the source I:1.1 = 200, then

$$N7:0 = \frac{200 \times 46}{10,000} + 50$$

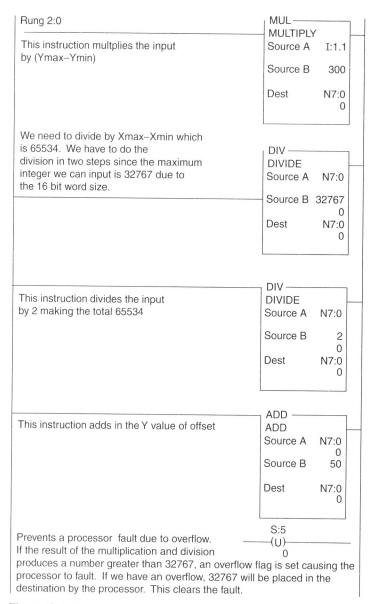

Rung 2:0

This instruction multplies the input
by (Ymax–Ymin)

MUL		
MULTIPLY		
Source A		I:1.1
Source B		300
Dest		N7:0
		0

We need to divide by Xmax–Xmin which
is 65534. We have to do the
division in two steps since the maximum
integer we can input is 32767 due to
the 16 bit word size.

DIV		
DIVIDE		
Source A		N7:0
Source B		32767
		0
Dest		N7:0
		0

This instruction divides the input
by 2 making the total 65534

DIV		
DIVIDE		
Source A		N7:0
Source B		2
		0
Dest		N7:0
		0

This instruction adds in the Y value of offset

ADD		
ADD		
Source A		N7:0
		0
Source B		50
Dest		N7:0
		0

S:5
——(U)——
0

Prevents a processor fault due to overflow.
If the result of the multiplication and division
produces a number greater than 32767, an overflow flag is set causing the
processor to fault. If we have an overflow, 32767 will be placed in the
destination by the processor. This clears the fault.

Figure 12.4 Sample program.

This instruction scales the channel 1 analog input value from an analog input module in slot 1 and stores the scaled value in N7:0.

Figure 12.6 does the same thing as Figure 12.4 and is also condensed into one instruction. It is available on the SLC 503 and above. It is called the Scale with Parameters (SCP) instruction, also based on $y = mx + b$. You need to know the input

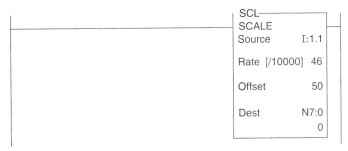

Figure 12.5   SLC 502 program equivalent to Figure 12.4.

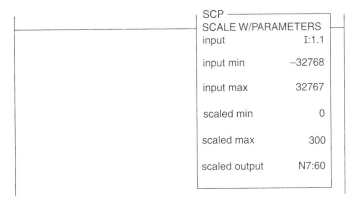

Figure 12.6   SLC 503 program equivalent to Figure 12.4.

and scaled maximum and minimums. The scaled value is the y-value. Again, the SLC 500 NIO4V analog I/O module, which has a full-scale integer range of $-32,768$ to $32,767$, is used to read the input. The analog input module provides the processor with a 16-bit binary number via I:1.1.

This instruction scales the channel 1 analog input value from an analog input module in slot 1 and stores the scaled value in N7:60.

► 12.2 PID INSTRUCTION

The processor of a PLC can have an instruction set aside for automatic control, called a proportional integral differential control *(PID)* instruction. The PID output instruction uses closed-loop control to automatically keep a process at a set point. This instruction usually reads analog input data from transducers, processes this information through an algorithm, and then sends analog output signals to devices to control the process to a desired set point . This task makes the instruction one of the most complex available for PLCs. A typical process is shown in Figure 12.7.

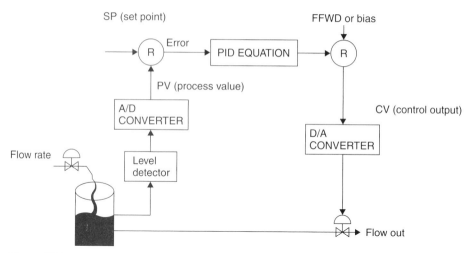

Figure 12.7 Typical PID loop.

Figure 12.7 shows a process in which you are trying to keep the level of the liquid in a tank constant. Refer to the PID loop in Figure 12.7 and consider the three types of control available.

12.2.1 Control Proportional to the Error

Control proportional to the *error* is the simplest type of control because you just determine the difference between the set point (SP) and the feedback value (PV) from the device to determine the error. Once you have the error, you multiply it by a constant and then send out this product as a *control value* (CV). If the constant is positive, it is called positive feedback, and if the constant is negative, it is called negative feedback. Proportional error has one drawback: when disturbance to the system occurs, an offset from the desired set point results.

Consider the example of proportional control shown in Figure 12.8(b). A liquid is fed into a tank through a valve controlled by a float-lever system that gives feedback about the liquid level in the tank. The proportional gain constant is determined by the position of the fulcrum of the lever being used between the float and the valve. In Figure 12.8(b) the fulcrum is adjusted to keep the water at a specified set point when the valve controlling the outflow is halfway open. This control system works well as long as the system is not disturbed. The inflow pressure and the outflow must stay the same for this feedback to control the level of the liquid in the tank correctly.

The problem with proportional control occurs when a change, such as an increase in the opening of the outflow valve in Figure 12.8(c), affects the system. Because the fulcrum position has not changed, the gain of the feedback has not

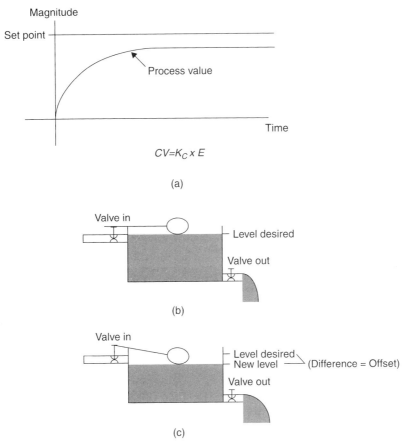

Figure 12.8 Proportional control; (a) Response of proportional control; (b) Original setting; (c) After a system disturbance.

changed, so the feedback is too low to maintain the set point. The result is an offset from the set point, which can be changed only by manually adjusting the fulcrum position to increase the proportional gain or by somehow adding more gain somehow automatically. When the system changes proportional control will result only in an offset. Integral control provides a way to eliminate this offset.

12.2.2 Control Proportional to the Integral

In control proportional to the integral, the error is integrated over time and multiplied by a constant. See Figure 12.9. This value becomes larger if the error stays positive as time goes on or until the error goes to zero and stays there. When the error is zero, the control value produced by this term remains constant until there is a disturbance in the system. The upper limit on the integral sign is a lowercase t,

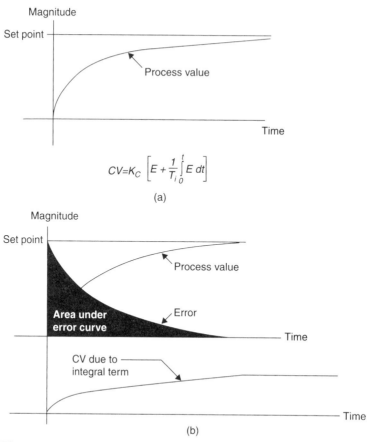

Figure 12.9 Integral control; (a) Response with P and I; (b) Variation of CV with error.

which means the integration will last as long as the PID function is active. This is an average value function only when $t = T$. When $t \gg T$, it becomes just the area under the curve multiplied by a gain of $1/T$. The integration goes on continuously and the error is usually changing. A new control value is produced by this term each time the PID instruction goes from false to true if an integer control block is used. Keep in mind the error can be positive or negative and can cause the error under the curve to increase or decrease. Each time the PID instruction goes from false to true, the PID algorithm calculates a new integral term, and adds it to the previous integral term, keeping a running total of the area under the curve. It does this mathematically by using the following dependent equation:

$$CV = K_c \left[\frac{1}{T} \int_0^t E \, dt \right]$$

The integration is done algebraically by keeping a running sum using the expression:

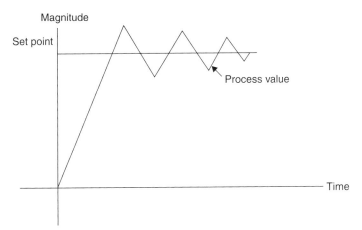

Figure 12.10 Derivative control.

$$CV_{new} = \frac{K_c}{T_i} \times E_{present} \times \Delta t + \frac{K_c}{T_i} \times E_{previous} \times \Delta t \qquad \text{where } \Delta t = \text{loop update time}$$

You determine Δt by the loop-update time you enter when using the PID instruction. The ladder logic determines when the PID instruction is going to be used and initiates the first execution of the PID algorithm. As long as the PID rung stays true, the algorithm is executed every scan. Executing the PID algorithm adds to *scan time,* so you do not want Δt faster than necessary. The best results are obtained when the loop-update time, Δt, is equal to the rate at which the PID changes from false to true. You will see in the program in Figure 12.12 that because the PV is updated just before the PID rung goes from false to true, the algorithm is executed by the CPU and, finally, the CV is updated. Having Δt faster or slower than the rate at which the PID instruction is toggled degrades the PID control significantly.

12.2.3 Control Proportional to the Derivative

In control proportional to the derivative, the derivative (rate of change) of the error, or the *process value,* is determined and multiplied by a constant. Fast changes result in large adjustments, and slow changes result in small adjustments. This function is seldom used alone because it tends to cause oscillations. However, using this function with P or I makes it possible to get to the set point faster. See Figure 12.10.

You can combine the error, integral, and derivative any way you like. One of the combinations commonly used is to use all three to determine the control value, as in Figure 12.11. Usually this combination will give you the best results and greater accuracy.

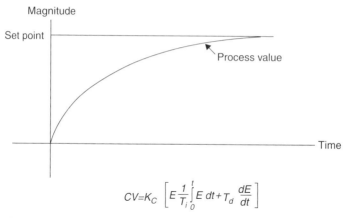

$$CV = K_C \left[E \frac{1}{T_i} \int_0^t E\, dt + T_d \frac{dE}{dt} \right]$$

Figure 12.11 Using PID.

You must make some decisions before you start to use the PID instruction. First consider what is required for a PLC 5. There are several standard equations used by manufacturers, and you need to choose one of the six equations to use for the PID instruction. Any one of these equations will produce the desired results. They are just different ways of manipulating the constants that control the effects of the three types of control.

There are two sets of three equations. The first set uses the same gain constant for each weighting of P, I, and D. These equations are called dependent gain equations. The second set allows you to weight P, I, and D separately; hence, the equations in this set are called independent gain equations.

Each set can specify a different derivative term as follows:

1. Positive derivative with respect to time of the error
2. Positive derivative with respect to time of the process variable
3. Negative derivative with respect to time of the process variable

The variables used in the equation are defined as follows:

K_P = proportional gain (unitless)   SP = set point
K_i = integral gain (repeats/second)   PV = process variable
K_d = derivative gain (seconds)   error = (SP− PV) or (PV − SP)
     bias = feedforward or external bias
K_c = proportional gain (unitless)   CV = control variable
$1/T_i$ = *reset* gain (repeats/minute)
T_d = rate gain (minutes)

The standard equations with dependent gains (ISA standard) are as follows:

$$CV = K_c\left[E + \frac{1}{T_i}\int_0^t E\,dt + T_d\frac{dE}{dt}\right] + \text{bias}$$

$$CV = K_c\left[E + \frac{1}{T_i}\int_0^t E\,dt + T_d\frac{dPV}{dt}\right] + \text{bias} \qquad (E = PV - SP)$$

$$CV = K_c\left[E + \frac{1}{T_i}\int_0^t E\,dt + T_d\frac{dPV}{dt}\right] + \text{bias} \qquad (E = SP - PV)$$

The independent gain equations are as follows:

$$CV = k_pE + k_i\int_0^t E\,dt + k_d\frac{dE}{dt} + \text{bias}$$

$$CV = k_pE + k_i\int_0^t E\,dt + k_d\frac{dPV}{dt} + \text{bias} \qquad (E = PV - SP)$$

$$CV = k_pE + k_i\int_0^t E\,dt - k_d\frac{dPV}{dt} + \text{bias} \qquad (E = SP - PV)$$

We can get the same results from either set of equations by making the gain constants equivalent:

$$k_p = K_c \qquad k_i = K_c\frac{1}{T_i\text{ min}} \times \frac{1\text{ min}}{60\text{ s}} = \frac{K_c}{60 \times T_i} \times \frac{1}{\text{s}}$$

$$k_d = K_cT_d\text{ min} \times \frac{60\text{ s}}{\text{min}} = 60K_cT_d\ (\text{s})$$

The PLC-5 requires you to decide to use integer or floating-point mathematics in your calculations. Floating-point will give you higher resolution, but floating-point is only supported by the most advanced processors. You select integer or floating-point by specifying the type of file, Nxx:x or PDxx:x, for the control block when entering the PID instruction. The N file causes integer mathematics and PD causes floating-point.

The following is the information for integer control. When you use the PID instruction, you will need to provide the following control-block information. A question mark means you have to provide the data and xx means you will see this information when you monitor the control block.

Integer data file control-block information (N7:xx) for the PID instruction is as follows:

Equation: ? 0:AB/1:ISA (0:independent/1:dependent)
Mode: ? 0:auto/manual

Error:	? 0: SP − PV/1:PV − SV
Output limiting:	? 0:no/1:yes
Set output mode:	? 0:no/1:yes
Set-point scaling:	? 0:no/1:yes
Derivative input:	? 0:PV/1:error
Last-state resume:	? 0:no/1:yes
Feed forward:	? xx
Max scaled input:	? xx
Min scaled input:	? xx
Deadband:	? xx
Set output value %:	? xx
Upper CV limit %:	? xx
Lower CV limit %:	? xx
Scaled PV value%:	xx
Scaled error%:	xx
Current CV %:	xx
Dead-band status:	x
Upper CV limit alarm:	x
Lower CV limit alarm:	x
Set point out of range:	x
PID done:	x
PID enable:	x
Set point:	? xx
Proportional gain (k_c):	? xx (unitless)
Reset time: (T_i)	? xx (min/repeat)
Derivative rate (T_d):	? xx (min)
Loop-update time:	? xx (s)

There are seventeen places that you either let default to zero or at which you provide information.

Equation AB/ISA (independent:dependent). Choosing independent basically lets you individually adjust the proportional integral and differential control. Choosing dependent applies a gain constant to all three simultaneously.

Mode: auto/manual. You can set the mode to either manual or auto. In the manual mode you control the output from a manual control station. You use the manual for determining tuning constants. (How this is done is discussed later in the chapter.) Auto lets the PID instruction control the output value.

Error: (SP − PV)/:(PV − SV). This setting determines if you want positive or negative feedback. You will know if you get this wrong on your first trial run because the wrong feedback will cause the system to be unstable, which shows up as a constantly increasing CV. You must be careful to limit your output if you are not sure which feedback to use to prevent damage to equipment or create an unsafe situation for personnel.

Output limiting.   You may want to limit the output in order not to put undo stress on your system.

Set output mode.   This setting determines if the control value is generated by the PID or inserted manually. Data is entered in the set output value word. For instance, you may want to use set output mode when you are trying to determine the correct constants for the PID equations. This process is called tuning, and one method requires a step input, which goes from full on to off. Tuning is discussed later.

Derivative input.   You can get the derivative of either the PV or error.

Set-point scaling.   You may display the set point in binary or engineering units.

Last-state resume.   When you need to make program adjustments and change the processor from the run to program mode and back, you may want the control output to hold its last state. The processor can save the integral accumulator and use it when you go from the program to the run mode again. This will keep the PID from making too large a correction when you initially go back to the run mode.

Feed forward.   Feed forward is often used to control processes with a transportation lag, as when heating a swimming pool. You first put the heater on high because you know it will take a while—perhaps many hours—for the water in the pool to heat up. However, if you leave the heaters on high, the water temperature may become much hotter than desired. If you want a PID to keep the water at just the right temperature, you have to turn off the feed forward so it can regulate the temperature properly. The feed forward usually works in conjunction with a timer and can be turned on and off by the ladder logic.

Max scaled input, min scaled input.   The PID displays various engineering-unit values on your monitor if you are running online and monitoring the program. The PID changes machine information in digital form to values humans can read, but to do this the program must have maximum and minimum values from transducers to scale the digital information to decimal values.

Deadband.   *Deadband* is a range of error in which the processor does not execute the PID instruction. This function keeps the CV value from hunting when the error gets small. Hunting is caused by calculating a value that makes the PV go too high, then compensating on the next calculation by making the CV lower, and then going back and forth. Deadband lets you set an acceptable range of error.

Set output value. This value is used for set output control of the CV output.

Upper CV limit alarm % and lower CV limit alarm %. These entries, given as percents, enable you to operate an alarm if the control output goes above or below specified values. If you do not want alarms, enter 0% for the lower limit and 100% for the higher limit.

Upper CV limit % and lower CV limit %. These entries, given as percents, enable you to prevent the control output from going above or below specified values. The upper limit can be used to control *antireset windup*. During startup the integral control can cause the CV to become too large, causing an *overshoot*. The error during startup is usually large, so the integral control rapidly increases its contribution during this period. The integral term can be turned off for a short period; during this period it is referred to as antireset windup. When the CV reaches its upper limit, which you set with output limiting, the integral control is automatically turned off. This eliminates the problem. You could call this anti-integral windup.

The terms *integral* and *reset* mean the same thing. When only proportional control was used because an offset from the set point always occurred, you would add a bias manual to bring the error to zero. This was referred to as resetting. However, integral control does this automatically. Those using only proportional control would say the integral term gives automatic resetting.

The lower limit can be used to add a bias. It can be added to the CV if you know there is a steady-state loss that has to be compensated. You may need a minimum amount of control value just to overcome friction. You can stop the need for correction at lower levels by adding a bias so that the CV value never goes lower than the amount needed to overcome this friction.

If you assume the dependent equation has been selected, you enter the following:

Set point	Enter the desired set point.
Proportional gain	Enter the proportional gain constant (unitless).
Reset time	Enter the integral gain constant (minutes/repeat).
Derivative rate	Enter the derivative gain constant (minutes).
Loop-update time	(s)

It takes time for the CPU to execute the PID algorithm, which affects the scan time. Therefore, you do not want it to update faster than is required by your system. The CPU updates the algorithm at the rate you specify whenever the rung containing the PID instruction is true. The CPU calculates a new CV value whenever the rung containing the PID instruction goes from false to true. After that, the next time the CV changes is determined by the loop-update time if the rung stays true. If the PID rung goes false, the CV stays at its last value. The loop-update time has to be coordinated with the rate at which you are sampling the PV and sending out the CV through your analog in and out. You must sample at the same rate or faster as the loop-update time for the PID control to work at its optimum.

When making data entries, be sure the units of the number you are putting in match those for which the data entry sheet is asking.

12.2.4 Sample Programs for Implementing PID Using a PLC-5

The first PLC-5 sample program (Figure 12.12) uses a timer to control the frequency with which the PID instruction sends out a new control value. It assumes the analog input and output modules are in the same chassis. In this example, the first rung sets up an interval timer for sampling the PV, and the second rung reads the analog input module through a block-transfer-read instruction as determined by the timer. The third rung turns on a bit that causes the PID instruction see a true only when the block-transfer read is completed. This is necessary because the block-transfer read is updated asynchronously during the main scanning. Because it is asynchronous, the block-transfer-read DN bit could go false before the scan is complete. The B3:0/0 will hold for one scan once it is turned on by the DN bit. The timer is used to control how often PV is read and PID sends out a new CV. This method is used for automatic-control loops in which the response of the process value to a change of the control value is slow. The loop-update time should be set at the same rate as the timer DN bit is cycling on and is determined by multiplying the preset by the time base.

The second example (Figure 12.13) executes the PID instruction every time new data are available from the analog input module. Because this occurs at a set rate that an input module samples data, we call this real time sample (RTS)-based. The RTS rate is set in the input module and the input module will not allow the BTR to initiate until new data is available. The PID is initiated by the BTR's done bit. The PID Loop Update Time is set to equal the RTS interval. The CPU executes the PID algorithm every time B3:0/0 goes high.

The third PLC-5 example, (Figure 12.14) is an interrupt subroutine that is called by a selectable time interrupt *(STI)*. The subroutine file number is located in word S:31. The frequency with which the STI subroutine is called is stored in S:30. The STI allows the user to interrupt the scan on a periodic basis, and the time interval stored in S:30 determines how often the PID instruction is executed. The loop-update time should equal the STI interval.

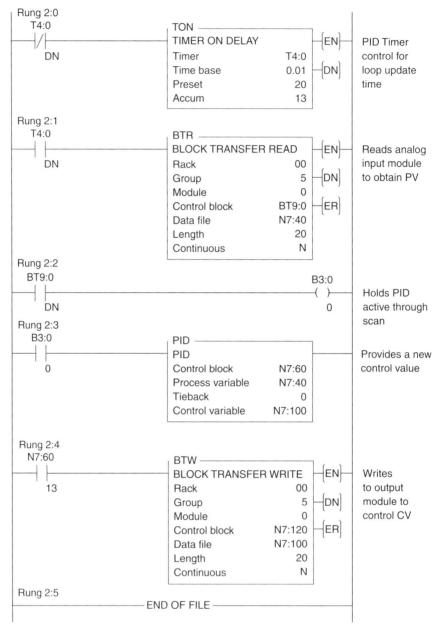

Figure 12.12 Using a timer to control the PID instruction.

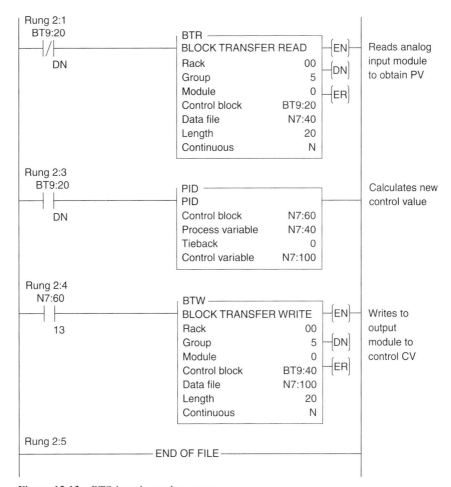

Figure 12.13 RTS-based sample program.

Processors at the SLC 502 level and above support the PID instruction, but they use only integer and not floating-point. The equation is

$$\mathrm{CV} = K_c \left[E + \frac{1}{T_i} \int_0^t E\, dt + T_d \frac{d\mathrm{PV}}{dt} \right] + \mathrm{bias}$$

This instruction requires you to fill in the data via a data-entry page that comes up when you insert this instruction. Some of these entries are the same as for the PLC-5. This information goes into a control-block file that is 23 words long, and the control block for the PID stores this information as follows:

```
Word   15 14 13 12 11 10 9 8  7  6  5  4   3  2  1  0
1      EN    DN PV SP LL UL DB DA TF SC RG  OL CM AM TM
2      PID Error Code (MSbyte)
```

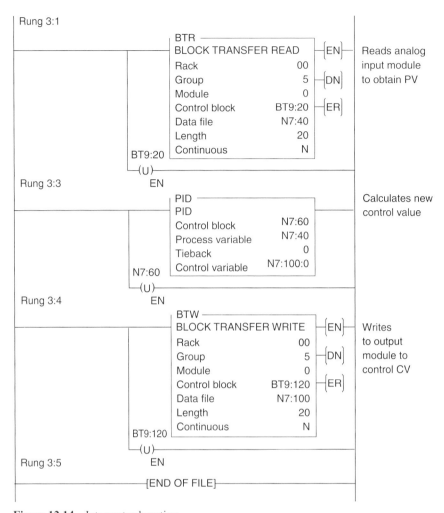

Figure 12.14 Interrupt subroutine.

3	Set point SP
4	Gain K_c
5	Reset T_i
6	Rate T_d
7	Feed Forward Bias
8	Set point Max (Smax)
9	Deadband
10	Internal Use—do not change
11	Output Max
12	Output Min
13	Loop Update
14	Scaled Process Variable

```
15    Scaled Error SE
16    Output CV% (0-100%)
17    MSW Integral Sum    503 MSW
18    LSW Integral Sum    503 LSW
19-22 Internal Use—do not change
```

The file should be an integer file. Be careful not to write into this file with other instructions elsewhere in your programming. The entries required for PID instruction are as follows:

```
Auto manual
Mode
Control
Set point(SP)
Process (PV)
Scaled error
Deadband
Output (CV)
Loop-update (0.01 s)
Gain (/100) (unitless) $K_c$
Reset (/100 min/repeat) $T$
Rate (/100 min)
Min scaled
Max scaled
Output (CV) limit
Output(CV) minimum
Output(CV) maximum
```

A brief description of each of these titles is given next.

Auto Manual.   You may set either manual or auto. In the manual mode you control the output by entering values manually through software while working online. You use manual for determining tuning constants. (How this is done is discussed later in the chapter.) The auto setting lets the PID instruction control the output value.

Mode.   The mode may be either timed or STI. The timed mode lets the PID instruction calculate a new CV output at a constant interval, which is set by the loop-update entry. The SLC 500 series PID instructions operate differently than the PLC-5 using an integer control block. The SLC 500 series PID instruction leaves the output term in its last state and clears the integral term when the rung goes false. Normally, the PID rung is unconditional because you do not want the integral term to be cleared. The SLC 500 series PID instruction calculates a new CV as determined by the loop-update time entered instead of using a timer, as in PLC-5 program (Figure 12.12). This method keeps the integral term from being cleared.

STI requires an interrupt routine to define the update interval. The STI subroutine then causes the PID to be updated every time the STI is scanned.

Control. $E = \text{SP} - \text{PV}$ or $E = \text{PV} - \text{SP}$, which determines if you want positive feedback ($E = \text{PV} - \text{SP}$), which results in an increase in the CV value when the PV becomes greater than the SP, or negative feedback ($E = \text{SP} - \text{PV}$), which results in a decreasing CV when PV becomes greater than SP.

Set Point. You need to tell the processor the set point in engineering units.

Deadband. The deadband is a range of error in which the processor will not execute the PID instruction, which keeps the CV value from hunting when the error gets small. Hunting is caused by calculating a value that makes the PV go too high, then compensating on the next calculation by making the CV lower, and then going back and forth. Deadband lets you set a range of acceptable error.

Output (CV). The % output is displayed and automatically calculated by the PID instruction in the auto mode. You set this value only in the manual mode.

Loop Update. This value (in hundredths of a second) tells how often the PID algorithm is executed.

Gain. This setting asks you to specify the proportional gain constant.

Reset. This setting asks you to specify the integral gain constant in minutes/repeat.

Rate. This setting asks for the derivative gain constant as a per-minute value.

Min Scaled. This setting gives the minimum engineering units the process will produce.

Max Scaled. This setting gives the maximum engineering units the process will produce.

Output CV limit. You answer yes or no to indicate if you want to limit the output.

Output CV Minimum. If you use output limiting, you need to give a percentage value.

Output CV Maximum. If you use output limiting, you need to give a percentage value.

Note that the SLC 500 series PID instruction is an integer-only algorithm. In the sample programs that follow, the PID instruction is in an unconditional rung.

Figure 12.17 uses an SLC 504 processor and an analog combination 1/O module to work with a PID instruction. The module is in slot 2, the address of the input channel being used is I:2.0, and the output channel is O:2.0. The scale-data instruction on the first rung reads the analog module in slot 2, word 0, which is one of the two input channels available on this module. It multiplies by the rate per 10,000 (slope) and then adds the offset. This scales the analog input information so it is compatible with the PID instruction. Figure 12.15 shows the graphic results of the calculation of the rate and offset. The input module receives a signal between 4 and 20 mA from a transducer and converts it to a value between 3277 and 16,384. Scaling this to a value from 0 to 16,384 makes it compatible with the PID instruction.

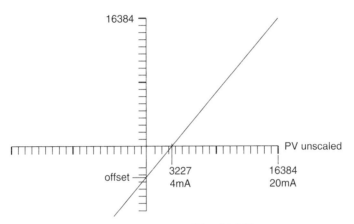

Figure 12.15   PV scaled to be compatible with PID range.

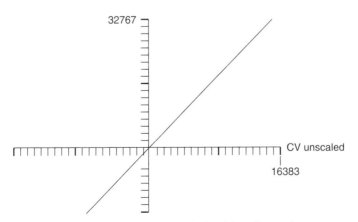

Figure 12.16   CV scaled to be compatible with analog out input range.

Using $y = mx + b$,

$$m = \frac{y_{max} - y_{min}}{x_{max} - x_{min}} = \frac{16,384 - 0}{16,384 - 3277} = 1.2499$$

$$\text{offset} = b = y - mx = 0 - 1.2499 \times 3277 = -4096$$

$$\text{rate}/10,000 = m \times 10,000 = 1.2499 \times 10,000 = 12,499$$

The second rung, which contains the PID function, calculates a new control value for the output module channel being used. It has an address of O:2.0. The third rung scales the CV value so it is compatible with the output module and then sends it out to the channel address as O:2.0

Figure 12.16 shows the graphic result of the calculation of the rate and offset so the CV is compatible with the output module. The output module receives an integer value from the PID and converts it to 0 to 10 V to manipulate a control value. This output module has an input range of 0 to 32,767. You need to scale the CV produced by the PID instruction to this range.

Using $y = mx + b$,

$$m = \frac{y_{max} - y_{min}}{x_{max} - x_{min}} = \frac{32,767 - 0}{16,384 - 0} = 1.9999$$

$$\text{offset} = b = y - mx = 0 - 1.9999 \times 0 = 0$$

$$\text{rate}/10,000 = m \times 10,000 = 1.9999 \times 10,000 = 19,999$$

The PID loop-update time is determined when you configure the PID; in this case you do it by setting the mode to timed mode and then entering the loop-update-time instruction. The analog I/O module is configured by hardware switches and the SCL both reads and scales the analog I/O. You must be careful when scaling data so the results do not produce values that are out of the range of the analog I/O or PID being used. If they are "out of range conversion," error will result.

You can add programming to check for out-of-range values and to turn on a bit so you know such values have occurred.

Figure 12.18 uses the same processor and I/O module but uses the STI for determining the loop-update time. The STI subroutine is shown in the figure. The IIM and IOM instruction ensure the analog I/O is updated just before and just after the PID is executed. Out-of-range detection instructions are used between the analog-in and the PID instruction. This checking may or may not be necessary and depends on the I/O module and processor being used. If the I/0 module cannot produce a value that is out of range for the processor or the processor cannot produce a value that is out of range for the I/O module, then this detection programming is unnecessary. Out-of-range detection may also be needed between the PID and the IOM.

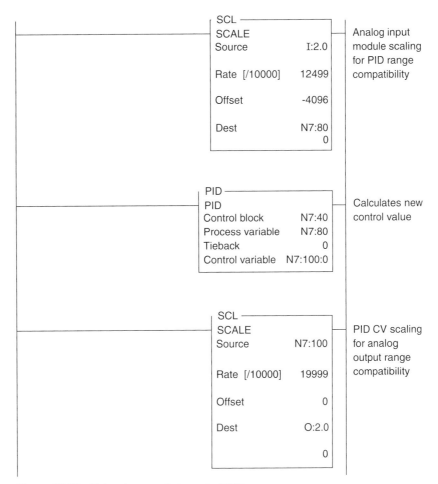

Figure 12.17 Using timer mode to control PID.

▶ 12.3 PID TUNING

You must *tune* the PID for it to work correctly. Improper tuning will result in severe oscillation or runaway, with the possibility of damage to personnel and equipment. You need to be very cautious during the trial runs and be able to stop the PID control if you make a mistake. Fast-acting systems present the most danger, but they are also easier to tune because you see the results faster.

Manufacturers use different equations for PID control, and it is important that you know which equation you are using. If you were to replace an existing PID controller with a PID from another manufacturer without realizing there is a difference in how the CV is calculated, you might try to use the same gains for P, I, and D used on the old controller. The new controller might then be unstable and not tuned.

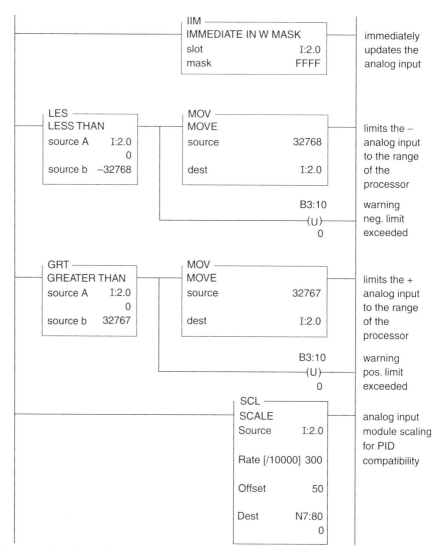

Figure 12.18 Using STI to control PID.

Figure 12.19(a) shows the two most popular equations for calculating the CV. The dependent equation is on the left, and the independent equation is on the right. If you solve the Laplace transform for the transfer function or gain in the s domain, you get the gain equations shown in Figure 12.19(b). Figure 12.19(c) shows the control-block diagram. Notice how the constants are equated so the gains in the s domain are equal. See Figure 12.11(d).

Different manufacturers may ask for these gains in different units, so you must watch the units carefully. Allen-Bradley uses the following:

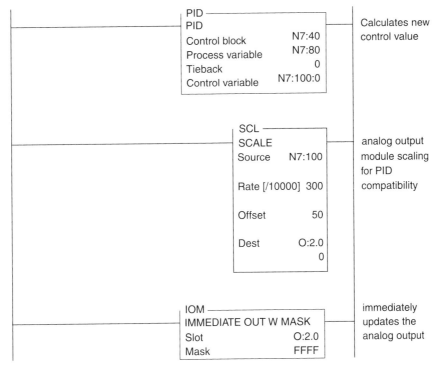

Figure 12.18 Continued

$$K_P = \text{proportional gain (unitless)}$$
$$K_c = \text{proportional gain (unitless)}$$
$$K_i = \text{integral gain (repeats/s)}$$
$$1/T_i = \text{reset gain (repeats/min)}$$
$$K_d = \text{derivative gain (s)}$$
$$T_d = \text{rate gain (min)}$$

For equivalence,

$$K_p = K_c$$
$$K_i \text{ repeat/s} = K_c/T_i \text{ (repeat/min)} \times (1 \text{ min}/60 \text{ s}) = K_c/60T_i \text{ repeat/s}$$
$$K_d = K_cT_d \text{ (min)} \times (60 \text{ s/min}) = 60K_cT_d \text{ s}$$

There is one more twist: you may need to give the PID gain entries per 100 or per 1000 because you may need decimal values, and this enables the numbers to be entered as integers. If you do not have a processor that can handle floating-point numbers, you have to enter the numbers as integers.

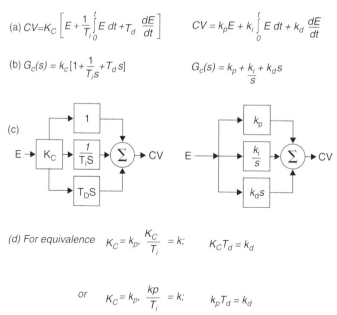

(a) $CV = K_C \left[E + \dfrac{1}{T_i} \displaystyle\int_0^t E\, dt + T_d \dfrac{dE}{dt} \right]$ $CV = k_p E + k_i \displaystyle\int_0^t E\, dt + k_d \dfrac{dE}{dt}$

(b) $G_c(s) = k_c \left[1 + \dfrac{1}{T_i s} + T_d s \right]$ $G_c(s) = k_p + \dfrac{k_i + k_d s}{s}$

(c)

(d) For equivalence $K_C = k_p,\ \dfrac{K_C}{T_i} = k;$ $K_C T_d = k_d$

or $K_C = k_p,\ \dfrac{kp}{T_i} = k;$ $k_p T_d = k_d$

Figure 12.19   (a) The two most popular equations for calculating the CV; (b) Gain equations; (c) Control-block diagram; (d) Constants equated for the gains in the s domain to be equal.

You may encounter other PID equations for calculating the CV. The instruction manual for your PID should give the proper equation. Remember if you are interchanging PID controllers, it is unlikely you can substitute the same gains and still have a system that is tuned.

The following three methods are used for tuning. Each has its own merits and like any tool, one may fit an application better than the other. These tuning methods will enable you to get to the point where you can fine-tune the system.

12.3.1 Direct Synthesis–Step Response Method

The direct synthesis–step response method is a sound method for tuning processes that relies on accurately determining the characteristics of a process by performing some simple tests.* First, you have to set up a control system similar to the one in Figure 12.20 and then turn the CV on in step fashion. This is called a bump test. Record the change in the PV. The constants for the PID equations are determined as follows.

---

*Kevin D. Starr, *Single Loop Control Methods*, Columbus, Ohio: ABB Industrial Systems Inc., 1996.

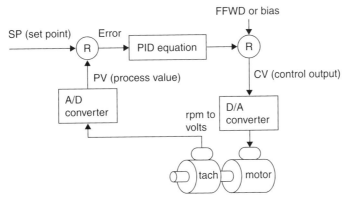

Figure 12.20 Control system.

These calculations assume we are using the standard equations with dependent gains:

$$CV = K_c\left[E + \frac{1}{T_i}\int_0^t E\, dt + T_d\frac{dE}{dt}\right]$$

The direct synthesis method is not good for determining the derivative constant, so it will be set to zero: $T_d = 0$.

First, do a bump test and record how the PV responds. A bump test is an open-loop test that is initiated by changing the CV instantly from one value to another and then observing the change in the PV. Figure 12.21 shows a bump test and a sample response. Record the percent change in CV and PV. Also record the dynamic settling time by observing how long it takes the PV to go to a new steady-state value.

Second, determine the first-order model confidence constant, or τ_{ratio}. How fast or slow do you want the process to respond? If you have a high level of confidence that the process will be stable and you want a fast response, pick $\tau_{ratio} = 1$; if you have a low level of confidence, pick the slow response: $\tau_{ratio} = 4$. See Figure 12.22.

Third, determine open-loop gain of the process and the open-loop time constant using the bump test results:

$$\text{gain of the process} = G_p = \frac{\%\ \text{change PV}}{\%\ \text{change CV}} = \frac{\dfrac{\text{change PV}}{\text{range PV}} \times 100}{\dfrac{\text{change CV}}{\text{range CV}} \times 100}$$

$$\text{open-loop time constant} = \frac{\text{dynamic settling time}}{4} = \Gamma_p$$

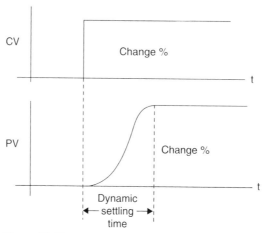

Figure 12.21 Bump test and sample response.

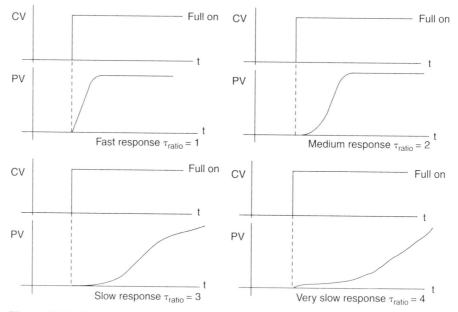

Figure 12.22 First-order confidence constant $\tau_{\text{ratio.}}$

Fourth, determine the controller gain and integral time.

$$K_c = \frac{1}{G_p \tau_{\text{ratio}}} \qquad T_i = \Gamma_p$$

12.3.2 Ziegler-Nichols Method Closed-Loop Test

A second method is the Ziegler-Nichols closed-loop test. This test was used for many years before the direct synthesis–step response method was developed. It is still used often, but the problem is that you must deliberately put the process in oscillation, and this can be a bit risky if not done carefully. Again, the equation for CV is assumed to be

$$CV = K_c \left[E + \frac{1}{T_i} \int_0^t E \, dt + T_d \frac{dE}{dt} \right]$$

First, set up the PID instruction for P action only by setting $T_i = \infty$ and $T_d = 0$. Set the PID to auto and the deadband to a wide range, and gradually increase the controller gain K_c until a steady oscillation occurs. Record the value of K_c and the period of the oscillations observed. See Figure 12.23. The constants are determined as follows:

$$K_c = 0.6 \times K_c \qquad \text{at which oscillation occurred}$$
$$T_i = 0.5T \, (\text{min})$$
$$T_d = 0.125T \, (\text{min})$$

You may not be able to use this tuning method if the driving force is not powerful enough to put the process into oscillation. The Ziegler-Nichols open-loop test is a better choice for slower-acting systems.

12.3.3 Ziegler-Nichols Open-Loop Test

The third method is the Ziegler-Nichols open-loop test, a method that has also been used by engineers for many years. First you have to set up a control system similar

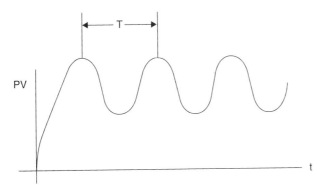

Figure 12.23   Oscillations observed.

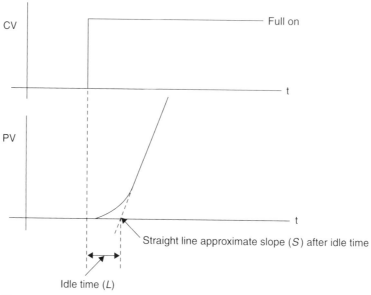

Figure 12.24 Determining the idle time.

to the one in Figure 12.20 and then turn the CV on as with a bump test. Record the change in the PV and then determine the slope when it becomes constant. It is constant when the curve traced by the PV approaches a straight line. (See Figure 12.24.) Project the straight line down until it meets the time axis. The idle time (*L*) is determined by measuring the difference in time from the intercept of the straight line and the point at which the CV is turned full on. Again, the equation for CV is assumed to be

$$CV = K_c\left[E + \frac{1}{T_i}\int_0^t E\,dt + T_d\frac{dE}{dt}\right]$$

The constants are determined as follows:

$$K_c = 1.2/LS$$
$$T_i = 2L\,(\text{min})$$
$$T_d = 0.5L\,(\text{min})$$

You may not be able to use this tuning method if the driving force is powerful and the process responds too fast because it causes an idle time that is so small it is unreadable. The Ziegler-Nichols closed-loop test is a better choice for fast-acting systems.

SUMMARY

Modern programmable controllers have made PID automatic control much easier to use than in the past. It is, however, one of the most complex PLC instructions, and learning correct data entry and tuning is critical because otherwise you can damage equipment or cause hazardous conditions for personnel. PID control also requires using A/D and D/A modules and learning how they interface with the PID instruction. It is important to match the units required by the PID instruction and the A/D and D/A modules. Once you have mastered these points, the PID control becomes a valuable tool for implementing automatic control.

▶ ## EXERCISES

1. A 10-bit analog input module is receiving a 0- to 10-V signal proportional to the speed of a motor. The most significant bit is the sign bit. Negative numbers are stored in 2's complement form, 0 is positive, and 1 is negative. The minimum speed of the motor is 0 rpm and the maximum is 3600 rpm.
   a. Determine the maximum negative and positive binary range of the input module.
   b. If the speed is 2500 rpm, what binary number will the input module send to the processor?
   c. If the binary word stored for an input is 11 0000 1111$_2$, what is the speed?
2. A 10-bit analog input module is receiving a 4- to 20-mA signal proportional to the temperature of a process. The most significant bit is the sign bit. Negative numbers are stored in 2's complement form, 0 is positive, and 1 is negative. The minimum temperature is 10°C and the maximum is 80°C.
   a. Determine the maximum negative and positive binary range of the input module.
   b. If the temperature is 50°C, what binary number will the input module send to the processor?
   c. If the binary word stored for an input is 00 0000 1111$_2$, what is the temperature?
3. We are connecting an analog input module with a range of +32,767 to −32,768 to a PLC with a PID instruction that has a range of +4095 to −4096. What scaled value needs to be sent to PID if the analog input module is reading 7575?
4. Write three different programs for scaling the analog-input module value going to the PID instruction. Use SCL for one program, SCP for one program, and neither of these for the other. This instruction will scale the channel 1

analog input value from an analog input module in slot 1 and store the scaled value in N7:50 for PID instruction.

$$Y \max = 16,383, \quad Y \min = -16,384, \quad X \max = 4095, \quad X \min = -1025$$

5. A bump test and record of the response are given in Figure 12.25. Determine K_c and T_i using direct synthesis.

$$CV = K_c \left[E + \frac{1}{T_i} \int_0^t E \, dt + T_d \frac{dE}{dt} \right]$$

6. The response of the PV is shown in Figure 12.26. An oscillation occurred at a K_c of 3 and the period of oscillation is $T = 1.3$ s. Determine K_c, T_i, and T_d for the equation $CV = K_c \left[E + \frac{1}{T_i} \int_0^t E \, dt + T_d \frac{dE}{dt} \right]$ by using the Ziegler-Nichols closed-loop method.

7. A bump test and record of the response are given in Figure 12.27. Use the equation $CV = K_c \left[E + \frac{1}{T_i} \int_0^t E \, dt + T_d \frac{dE}{dt} \right]$, determine K_c, T_i, and T_d by using the Ziegler-Nichols open-loop test.

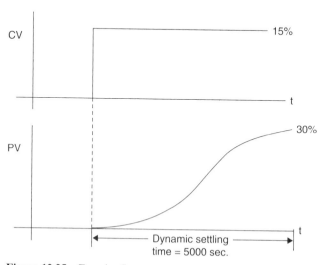

Figure 12.25   Exercise 5.

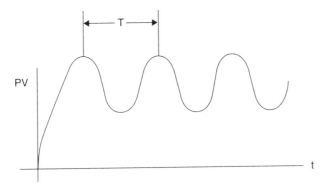

Figure 12.26 Exercise 6.

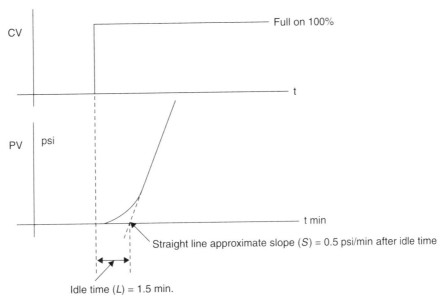

Figure 12.27 Exercise 7.

8. What would K_p, K_i, and K_d be if you used the information in Exercise 7 with the following independent equation:

$$\text{CV} = K_p E + K_i \int_0^t E\, dt + K_d \frac{dE}{dt}$$

9. Which of the three types of control (P, I, or D) is seldom used by itself? Why?

10. A 100-ft-tall display thermometer has been built for a local town square, and an SLC 500 PLC controls its operation. See Figure 12.28. There will be a reservoir containing red water and antifreeze that will be pumped up into a glass tube with graduation marks to give a visual display of the temperature. Two transducers are used, one that measures the temperature of the outside air and the other that measures the pressure in the reservoir due to the column of water. These transducers go to channel 1 and 2 of a 16-bit analog input module. There is also and electric valve to let the fluid out of the glass tube if the fluid needs to be lowered. The control adjusts the column of fluid to match the outside-air reading of the temperature transducer. The temperature transducer is connected to channel 1 and has an output range of 4 mA to 20 mA; the pressure transducer is connected to channel 2 and has an output range of ± 10 V. There is a linear relationship between the height of the water column and the pressure. A test is conducted; when the column is at -100 then the pressure transducer is at 4 mA and when the column is at 200 then the pressure transducer is at 13 mA. The temperature transducer puts out 2 V at $-100°$ and 8 V at $200°$.

Your job is to determine the scaling for each transducer so that the temperature transducer can be compared to the reading produced by the water column to see if they match. Make a state diagram and a structured subroutine program to implement this control. The scaled reading from the temperature transducer is stored at N7:10 and the scaled reading from the pressure transducer is stored at N7:11. Determine the correct scaling before storing the data at locations N7:10 and N7:11.

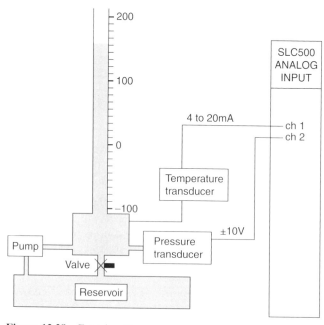

Figure 12.28 Exercise 10.

13 Programmable Controllers: Selection, Installation, Grounding and Safety, Troubleshooting, and Maintenance

---

CHAPTER OUTLINE

LEARNING OBJECTIVES

Upon reading this chapter students should be able to:

- Select a correct PLC and modules when given an application.
- Describe the correct grounding of a PLC for safety and for preventing interference.
- Draw a safe shutdown control scheme to be used when the processor or module fails.
- Describe what needs to be done to maintain a PLC in good operating condition.
- Troubleshoot the control and get it operating when given a nonoperating program.

INTRODUCTION

This chapter deals with actually selecting a programmable controller and installing, troubleshooting, and maintaining it. The hardest task is figuring out what to buy. It is difficult because there are many manufacturers and each manufacturer has many models. This chapter gives a checklist of features, and this list helps you consider all the options. Programmable controllers are user-friendly and are easy to install, troubleshoot, and maintain, but there are also safety issues to be considered. This chapter gives you the information you need to properly purchase, use, and maintain a PLC.

▶ 13.1 SELECTING A PROGRAMMABLE CONTROLLER

Selecting a programmable controller is similar to selecting a personal computer because there are so many manufacturers, models, and options. There are many details to consider, so it is easy to miss something that will turn out to be crucial for your application. Having a restricted budget and making a purchase that must last several years serve as an indication that you should take your time and do some homework. The best way to tackle the details is to make a check-off list. The list of questions is as follows.

1. What size PLC is needed?

Comments

a. Number of I/O points _____
b. Single task controlling one process _____
c. Multitask controlling several processes _____

2. What size memory is required?

Comments

a. Number of I/O points _____
b. Size of control program _____
c. Structured programming _____
d. Data-collecting requirements _____
e. Complex instructions _____

3. What type of architecture is needed, fixed or modular?

Comments

a. Fixed _____
b. Modular _____

4. What special functions are needed?

                                                                      Comments

    a. PID   _____

    b. Sequential function chart   _____

    c. Floating-point math   _____

5. What discrete I/O modules are needed?

                                                                        Voltage

    a. Current-sinking dc input module   _____

    b. Current-source dc input module   _____

    c. Fast-response dc sinking-input module   _____

    d. Current-sinking TTL input module   _____

    e. Current-source TTL input module   _____

    f. AC input module   _____

    g. AC output module   _____

    h. High-current ac output module   _____

    i. Current-sinking dc output module   _____

    j. Current-source dc output module   _____

    k. Current-sinking TTL output module   _____

    l. Current-source TTL output module   _____

    m. Relay output module   _____

    n. Combination input-output module   _____

    o. Electronically protected current-sourcing   _____
       output module

6. What analog I/O modules are needed?

                                                               Voltage or
                                                               current rating

    a. Analog input module   _____

    b. Analog output module   _____

    c. High-resolution analog input module   _____

    d. High-resolution analog output module   _____

    e. High-speed analog input module   _____

    f. High-speed output module   _____

    g. Number of input channels   _____

    h. Number of output channels   _____

7. What specialty modules are needed?

Comments

   a. Thermocouple   _____

   b. Barrel temperature   _____

   c. RTD   _____

   d. High-speed counter encoder   _____

   e. Open-loop velocity control   _____

   f. Synchronized axes   _____

   g. Mold pressure   _____

   h. BASIC language   _____

8. How will program memory be made nonvolatile?

Comments

   a. Battery-backed-up RAM   _____

   b. PROM   _____

   c. EPROM   _____

   d. EEPROM   _____

9. What are the communications requirements for the processor for networking and programming?

Comments

   a. DH 485   _____

   b. RS-232   _____

   c. PIC   _____

   d. SCADA   _____

   e. Ethernet   _____

   f. Device Net   _____

   g. DH+   _____

10. How will the PLC be programmed?

Comments

   a. Hand-held programmer   _____

   b. Portable programmer   _____

   c. Personal computer   _____

11. Is programming software needed?

Comments

   a. Manufacturer's   _____

   b. Generic   _____

12. What new processor or equipment is available?

                                                Comments

    a. Processor   _____

    b. Equipment   _____

13. Does the PLC have to be compatible with an older system?

                                                  Comments

    a. Yes   _____

    b. No   _____

14. Which manufacturer is best?

                                              Comments

    a. Products available   _____

    b. Training available   _____

    c. Stable company   _____

    d. Competitive price   _____

    e. Support   _____

15. Do special environmental requirements dictate the use of a special enclosure?

                                              Comments

    a. Excessive ambient temperature   _____

    b. Corrosive atmosphere   _____

    c. Dusty atmosphere   _____

    d. Tropical atmosphere   _____

    e. Excessive moisture or water   _____

    f. Vibration   _____

    g. Explosive or hazardous atmosphere   _____

16. What hardware accessories are needed?

                                              Comments

    a. Replacement part   _____

    b. PanelView   _____

    c. Bar-code reader   _____

    d. Others   _____

This section discusses each of these 16 questions and also gives some detailed information to help you make a decision.

1. What size PLC is needed?

	Comments
a. Number of I/O points	_____
b. Single task controlling one process	_____
c. Multitask controlling several processes	_____

Figure 13.1 shows how the size of a PLC is determined by the number of I/O points available. You must take into account both your present and future needs in regard to I/O. Determine how much you will need and add on a percentage for the future.

The application of a PLC also determines the size PLC needed. If it is single-ended and used to control only one process, this usually indicates a smaller size will be adequate. A carwash could easily be controlled by a nano, micro, or small PLC, and in this case the number of I/Os determines the size you need. A single process being controlled could go from simple to very complex tasks, and the complex tasks may require a medium to large PLC. Multitasking will usually require a medium to large PLC because the processor has to control several processes and the instruction set needs to have special instructions available to help with this process. Sequential Function Charts is Allen-Bradley's name for a Petri network, which is a form of structured programming. This type of structured programming is available only on larger machines.

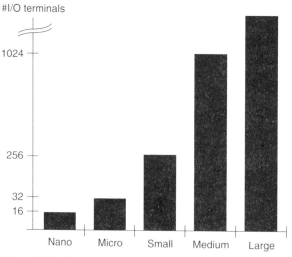

Figure 13.1 Determining the Size of a PLC by the Number of I/O Points.

2. What size memory is required?

Comments

   a. Number of I/O points   _____

   b. Size of control program   _____

   c. Structured programming   _____

   d. Data-collecting requirements   _____

   e. Complex instructions   _____

Determining the size of memory is important. There are several factors you need to consider. However, if you have little or no experience with programming, you may not be sure about how much memory you will need. In order to get help from people who sell PLCs, you must first do some preparation so that you know what you want. Determine answers to points a through e. For example, for c and d, you can at least answer yes or no. Each of these points affects the size of the memory you will require. The amount of memory taken up by each item is machine-dependent, so that makes it difficult to determine memory size.

Item a is the easiest to answer because you can count the I/O points and add a future amount. Item b can be estimated by counting the lines of programming you need if you have a simple process that does not require complex instructions. Ask a sales representative about the average memory required for a rung of instruction. Item c, structured programming, uses a lot of memory, but it has many advantages and you can review these by going back to Chapter 10. You may need to add as much as 25% more program memory for structured programming. Item d, data collecting, always requires extra memory because the information being collected has to be stored at least temporarily. The amount of additional memory needed depends on how much data you are collecting. Item e, complex instruction (such as PID control), uses significantly more memory than simple instructions. This is a case when you may have to go to the manufacturers and ask for estimates. Being prepared will help you get the right information.

3. What type of architecture is needed, fixed or modular?

Comments

   a. Fixed   _____

   b. Modular   _____

Fixed I/O is usually your only choice for nano and micro PLCs, whereas medium and large will usually require modular I/O. Small PLCs come in both varieties, and you will need to make a choice. Fixed I/O is less expensive to build and therefore costs less. However, fixed I/O is recommended only if you plan to have the PLC do a simple control process or one that you do not plan to change. If you have a fixed system and you need a different I/O, you will have problems if you want to change to a different process.

For example, you may buy a PLC with relay outputs and decide to move to a process that requires the outputs to change very fast. This situation requires triac- or transistor-driven outputs and thus makes the relay output useless. Hence, you would have to buy another PLC. Modular I/O, on the other hand, is more expensive but gives you much greater flexibility. The power supply, processor, and modules go into a chassis or rack design. You can change any of these items, so you can design your PLC around your application. If you need triac rather than relay outputs, you simply switch modules.

4. What special functions are needed?

Comments

a. PID _____

b. Sequential function chart _____

c. Floating-point math _____

You may need to add to this list if you have other special functions. PLC processors come in stripped-down versions and versions that have everything. You need to determine if the functions you want are available on the PLC you are buying.

5. What discrete I/O modules are needed?

Voltage

a. Current-sinking dc input module _____

b. Current-source dc input module _____

c. Fast-response dc sinking-input module _____

d. Current-sinking TTL input module _____

e. Current-source TTL input module _____

f. AC input module _____

g. AC output module _____

h. High-current ac output module _____

i. Current-sinking dc output module _____

j. Current-source dc output module _____

k. Current-sinking TTL output module _____

l. Current-source TTL output module _____

m. Relay output module _____

n. Combination input-output module _____

o. Electronically protected current-sourcing output module _____

Deciding on discrete I/O requires you to make decisions about four things. First, what voltage source is going to be used with the module? Second, will the connected circuits require sinking or sourcing? Third, is speed of operation a problem? Fourth, do you want the output to be self-protecting?

The common voltages are 24 V dc, 48 V dc, 120 V ac, 240 V ac, and TTL, which is 5 V dc. If you need different voltages for the output modules and you do not want to buy several different types of output modules, you can go to a relay output. The output can be a dry contact that has nothing connected to it, and you can apply any voltages you want provided you do not exceed the ratings of the contacts. Figure 13.2 shows that current sourcing and sinking depend on whether the source is coming from the module or from the device connected to the module. Current sourcing occurs when the plus terminal of the power source is connected to the module; the module then provides current to the device connected to it, as in Figure 13.2(a) and (c). Current sinking occurs when the plus terminal of the power source is connected to the device; the device then provides current to the module connected to it and the module provides a return path, as in Figure 13.2 (b) and (d).

If you need speed, you cannot use relay modules. You have to go to solid-state modules. A relay takes 10 to 20 ms to pick up as compared to a solid-state device, which can operate in microseconds. The one problem with the solid-state modules is that you can damage the solid-state output device if the output terminal is shorted. Fuses can help, but a full-blown short can result in current so high that the fuse is not fast enough. One solution is to go to electronically self-protected modules, which are fast and can prevent damage from shorts.

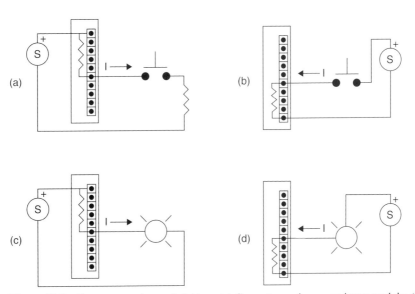

Figure 13.2 Current sourcing and sinking; (a) Current sourcing on an input module; (b) Current sinking on an input module; (c) Current sourcing on an output module; (d) Current sinking on an output module.

6. What analog I/O modules are needed?

		Voltage or current rating
a.	Analog input module	_____
b.	Analog output module	_____
c.	High-resolution analog input module	_____
d.	High-resolution analog output module	_____
e.	High-speed analog input module	_____
f.	High-speed output module	_____
g.	Number of input channels	_____
h.	Number of output channels	_____

To determine the analog I/O you need, you must answer the following: Will you will be working with current or voltage? What are the speed requirements? How many channels do you need? What resolution is necessary? The common analog input and output ratings are 0 to 1 V dc, +1 V dc to –1 V dc, 0 to 5 V dc, +5 V dc to –5 V dc, 0 to 10 V dc, +10 V dc to –10 V dc, and 4 to 20 mA. Your application will determine what you must use. Resolution is determined by how many bits are available in the module to store the analog information in digital form. If 16 bits are available, then the resolution of your selected voltage or current can be divided into $2^{15} = 32,768$ different increments, assuming the sign bit is the most significant bit. Different modules have different resolutions, so you need to look at the number of bits available for storage and your application requirements.

7. What specialty modules are needed?

		Comments
a.	Thermocouple	_____
b.	Barrel temperature	_____
c.	RTD	_____
d.	High-speed counter encoder	_____
e.	Open-loop velocity control	_____
f.	Synchronized axes	_____
g.	Mold pressure	_____
h.	BASIC language	_____

There are many other specialty modules available in addition to those listed here. The thermocouple module interfaces with types J, K, E, R, S, B, C, and D thermocouples, but not all modules have all these types. You need to check to see if the thermocouple you are using is covered by the module you buy. These modules enable you to read millivolt signals that standard analog modules cannot read.

They are designed to change the millivolt signal into scaled engineering units such as degrees Celsius or Fahrenheit and to work with PID instructions.

Barrel-temperature modules enable you to monitor four zones of autotuned PID heating or cooling for temperature control. Molding machines and extruders commonly use this module for controlling barrel temperature while injecting material. Make sure the processor you are using can support this module.

RTD stands for resistance temperature detector and is used to measure temperature injunction with platinum, nickel, copper, and nickel-iron variable-resistance devices. Similar to the thermocouple module, it is designed to change the resistance read into scaled engineering units such as degrees Celsius or Fahrenheit and to work with PID instructions.

High-speed counter-encoder modules enable you to count and encode faster than you can with a regular control program written on a PLC, where the speed at which the control program can be executed is too slow. It has the necessary electronics to count independently of the processor and to store the information or to send it out. The processor can access and manipulate its information through control programs. You need to check the frequency response of this module to make sure it is compatible with your needs. A common range available is from 0 to 50 KHz, which is able to count 50,000 pulses/s.

Open-loop velocity-control modules are used to control hydraulic presses, weld-head placement, and die-casting machines, controlling the velocity of an operation. For example, you might need to accelerate fast to start the operation of a press with a piston but then need to slow it down as the press process is being completed.

A synchronized-axes module provides logic to synchronize multiple axes. Examples of applications are hydraulic tailgate loaders, forging machines, and roll-positioning operations.

A mold-pressure module is used to connect with strain gauges for detecting mold cavity and hydraulic pressure. This information is used in a control program to improve repeatability.

The BASIC module lets you program your PLC in BASIC rather than in relay ladder logic.

You can add other modules to this list, and it is a good idea to check to see if special modules have been made for your particular application.

8. How will program memory be made nonvolatile?

Comments

a. Battery-backed-up RAM   _____

b. PROM   _____

c. EPROM   _____

d. EEPROM   _____

There are several ways of making sure you do not lose your program if a power outage occurs. The most common is battery-backed-up RAM. This system simply relies on a battery to maintain memory. The batteries are kept charged by a charging circuit in the PLC, which operates off of the power supply to the PLC. Batteries will hold the memory for long periods of time when they are fresh. They need to be checked periodically. Because RAM can be read and written to remotely over a network, other PLCs and computers can modify the PLC program memory.

PROM will hold a program forever, but if a change needs to be made, a new unit will have to be programmed and the old one will probably have to be thrown away. EPROM can be erased with ultraviolet light and reprogrammed if necessary, and the same EPROM can be used over and over. EEPROM is electrically erasable and programmable and may be what you want if a computer is going to be supervising your PLC. The computer can then change the program remotely.

9. What are the communications requirements for the processor for networking and programming?

	Comment
a. DH 485	_____
b. RS-232	_____
c. PIC	_____
d. SCADA	_____
e. Ethernet	_____
f. Device Net	_____
g. DH+	_____

You must consider your communications requirements carefully because you have many options. You need to determine how your personal computer is going to communicate with the processor to load, edit, save, and download control programs. You also need to know if your PLC is going to communicate with other processors or networks and if you will be doing supervisory control and data acquisition.

DH 485 is a connector and port for a local network connecting your PLCs. If you need to have a local network for your PLCs, then you will need to have a PLC with this connector. The communication between PLCs will take place automatically after you provide the configuration information asked for by the software used to program the PLC. You have to assign a name or number identification to each PLC and to the terminal used to program it. Once you have configured the system properly from one terminal, you can upload and download programs and talk to any PLC on the network.

An RS-232 connector and port are needed for direct connection to an IBM-compatible computer-programming device, for SCADA communications, and for connections to ASCII devices such as weigh scales, printers, and bar-code readers.

SCADA stands for supervisory control and data acquisition, which is used to categorize the software and systems that many companies have for accomplishing this task. You can obtain information about available SCADA software from any sales office. Rockwell International's RS View 32 enables you to do data acquisition, data collection, alarm management, advisory control, data monitoring, manufacturing of execution systems, and statistical process control directly via your PLC.

PIC stands for personal computer interface connector. A PIC is used for communication between a communication port on your personal computer and the PLC. The older PLCs required that you purchase a communication card and install it inside your personal computer, but later PLCs let you use the communication ports available on a PC without having to purchase an internal communication card.

An Ethernet connector and port enable you to connect to a high-speed network (10 megabits/s). Ethernet is a *protocol* that determines the physical arrangement of the connector and the way communication will take place. There are three standard connectors: AUI (asynchronous universal interface DB 15 connector), BNC (bannet connector), and UTP (unshielded twisted pair). A processor with an Ethernet connector will have the hardware and programming to automatically handle Ethernet communication.

Device Net is fast becoming a popular way to connect devices to a PLC. Rather than hardwire a device to a PC, the device is connected to a network called Device Net. You must have a smart device, which has a microprocessor built into it to enable the device to communicate on the network. They can save money by reducing hard wiring, plus by having the microprocessor on the device, you can control the device by having the PLC send commands over a network. Data collection is also available.

DH+ is a local information network that is designed for remote programming and for accessing and transferring of data. It is the primary communications network for the SLC 5/04 and PLC-5® controllers. There is a maximum of 64 nodes on a local network, numbered 0–77 octal. There are three different baud rates available on DH+: 57.6k baud (maximum cable length, 10,000 ft), 115.2k baud (maximum cable length, 5000 ft), and 230.4k baud (maximum distance, 2500 ft).

10. How will the PLC be programmed?

Comments

    a. Hand-held programmer   _____

    b. Portable programmer   _____

    b. Personal computer   _____

There are several ways you can program a PLC. A hand-held programmer is the cheapest, but it has limitations. You must put in the program one step at a time. You will have to write the programming steps down on a piece of paper and then use a keyboard to enter these instructions. It is not good for troubleshooting because you cannot watch the rung of logic being executed as you can when programming with your personal computer and the required PLC software. It can be handy for making limited changes because it is small and easy to take from one location to another.

Some manufacturers have made portable programming devices with LCD screens that are like portable computers, but they can interact only with a particular PLC. You can see several rungs of control, which is nice for troubleshooting.

It is hard to beat programming with your personal computer because you can do everything that a hand-held or a portable programmer can do plus you can make and modify good documentation of your program. You probably already have a personal computer; all you need is the software and a way to get your computer to communicate with your PC. Manufacturers communicate in various ways to PCs, and different models from the same manufacturer use different methods. Sometimes you will use a card that goes in your PC, or you may go from your serial port to a PIC (personal computer interface connector) to the PLC. Question 9 asks about communications for programming.

11. Is programming software needed?

Comments

    a. Manufacturer's   _____

    b. Generic   _____

Software for programming PLCs is always available from the manufacturer, and this software should work the best because the coordination of creating and testing software is best done in-house. This software usually can operate only on the processor of its manufacturer. Generic software is available that works with the PLCs of several manufacturers, and it tends to be more economical. Obviously, there may be some compatibility problems. If you look at trade journals, magazines, and websites, you can learn about the generic software available.

12. What new processor or equipment is available?

Comments

    a. Processor   _____

    b. Equipment   _____

Automation technology is constantly changing and evolving, just as computers are. The computer or PLC you are using today will be replaced by a newer, better, and cheaper one within a few months after you buy it. It is expensive to keep up with the latest equipment, and generally you have to live with what you buy for some time. You need to look at trade journals, magazines, and websites of PLC manufacturers to see what they are planning and what new equipment is available.

13. Does the PLC have to be compatible with an older system?

                                                                    Comments

    a. Yes   _____

    b. No   _____

You tie a substantial amount of money and engineering to a particular PLC manufacturer or model over time. This can make compatibility a major issue. Will the new PLC work with all the existing equipment? Seldom can you throw out all your old equipment and put in all new. Training may be an issue here too. If you go with a new system, will you have to retrain personnel, and would it be economical to do so? Maintaining and troubleshooting two different systems could also be a problem.

14. Which manufacturer is best?

                                                                    Comments

    a. Products available   _____

    b. Training available   _____

    c. Stable company   _____

    d. Competitive price   _____

    e. Support   _____

There are many manufacturers of PLCs, and each manufacturer has many modules. If you fill out this questionnaire and give it to a PLC vendor, the questionnaire becomes a specification for what you need. Some manufacturers may be eliminated because they do not have a module or a particular feature you need. Training may be an important consideration; many manufacturers offer training at any time at dozens of locations throughout the country.

The competition among PLC manufacturers is fierce, and some manufacturers may not survive, thus potentially leaving you with a PLC with no support. This is not likely to happen with the big companies such as Allen-Bradley, General Electric, or Siemens. The bigger companies with good reputations are able to charge more for their PLCs than smaller companies who are trying to break into the market or to fill a niche that is not covered by the larger companies. You may find a competitively priced unit that has the same features as a more expensive model but costs less.

Another issue is support. If you have a problem, can you easily find the technical support you need? If you have a problem with a critical PLC on an assembly line that shuts down the line, it may cost several thousand dollars an hour. In such a case you need to be able to get good support fast.

15. Do special environmental requirements dictate the use of a special enclosure?

	Comments
a. Excessive ambient temperature	_____
b. Corrosive atmosphere	_____
c. Dusty atmosphere	_____
d. Tropical atmosphere	_____
e. Excessive moisture or water	_____
f. Vibration	_____
g. Explosive or hazardous atmosphere	_____

The environment in which the PLC will be placed must be taken into account. The presence of any item listed here means the PLC must be put in a protective enclosure that will eliminate the problem created by the condition. Standard enclosures are readily available from manufacturers, and *NEMA* (National Electric Manufacturers' Association) *standards* have been written for testing and manufacturing these enclosures. It is usually more economical to buy one of these standard enclosures than to make one yourself. It may be economical to put the PLC in a remote location and use remote I/O; however, the remote I/O will need protection.

16. What hardware accessories are needed?

	Comments
a. Replacement part	_____
b. PanelView	_____
c. Bar-code reader	_____
d. Others	_____

Manufacturers make equipment that can be added to a PLC to give it greater functionality, and you may want to purchase some of these items.

The replacement part you need depends on what you have already chosen to buy in the other 15 questions. Do you want to have a spare processor or module in case you have a problem? You can damage equipment by doing something such as shorting an output. How critical is the process you are controlling? This will determine if you want to have spare modules, etc.

A typical helpful accessory is a human-machine interface such as PanelView. Figure 13.3 shows a picture of three different grades of PanelView available from Allen-Bradley.

Another good accessory is a bar-code reader, shown in Figure 13.4

If you are a neophyte, a good salesperson can help you make the right decisions. Your part is to educate yourself by using all the resources you can find. Keep in mind that PLCs are constantly evolving to adapt to new technologies and to

(a)

Figure 13.3 PanelView interfaces; (a) PanelView 900; (b) PanelView 1200E; and (c) PanelView 1400E. Courtesy of Rockwell Automation/Allen-Bradley.

maintain competitiveness. Therefore, the criteria used to select a PLC will also change. The questions asked in this chapter will need to be revised, but they can be used as the basis for a new list. Selecting a modern PLC is like selecting a modern PC—not a simple task because of all the options that have evolved. Selecting an overpowered PLC on the one hand or an inadequate one on the other can be a costly mistake.

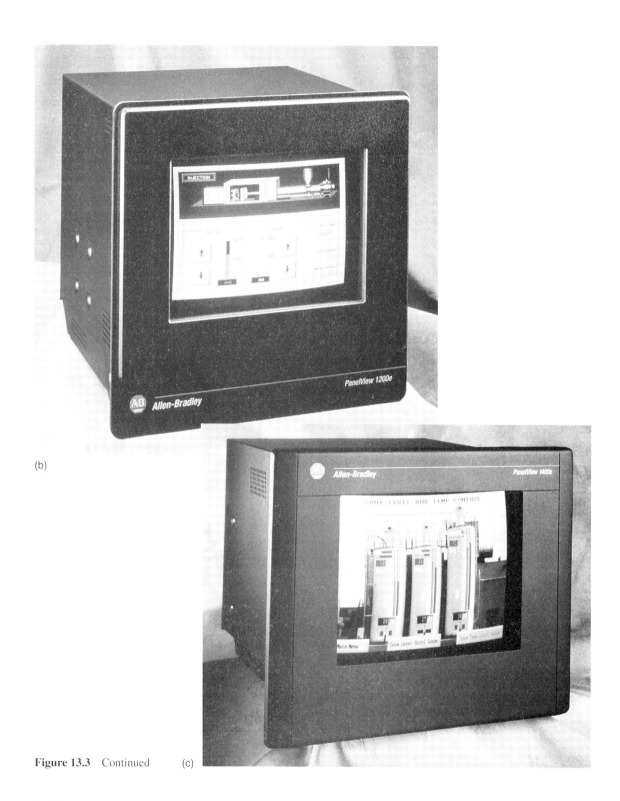

(b)

Figure 13.3 Continued (c)

Figure 13.4 Bar-code reader. Courtesy of Rockwell Automation/Allen-Bradley.

▶ 13.2 INSTALLATION, GROUNDING, AND SAFETY

Installing a programmable controller requires several considerations in order for it to operate properly and safely. You need to note the cautions in the installation manual for the particular PLC you are installing.

Selecting a location for the controller is usually the first decision that has to be made. It is usually safer and easier to troubleshoot control programming if the device you are controlling is near the PLC. This is not always practical if you are controlling a process that physically covers a large area, such as a paper mill or a chemical-processing plant. Then a centrally located PLC with remote I/O makes sense. The space around the controller should be large enough to enable the people who have to make connections to the modules to be able to work comfortably. Usually a minimum of 3 ft immediately in front of the PLC is required to work comfortably. OSHA requirements also have to be considered. You must provide necessary network connections for the PLC if the processor is to be networked. Different connectors and network drops are required. Make sure the PLC is in a location that can be made accessible for the network cabling required.

As stated previously, the PLC's environment may dictate that it has to be in an enclosure. We mentioned dust, wetness, and vibration, for example, as such environments. The enclosure may affect your location selection and almost certainly will affect your work space. Make sure you have room to open doors and work effectively.

Grounding needs to be done correctly for proper safety and for preventing interference on incoming and outgoing electrical signals. Figure 13.5(a) shows a properly grounded programmable controller. There is a neutral-ground bus and an equipment-ground bus in a power-distribution panel. The neutral provides the return for the power circuit. The equipment-ground bus keeps equipment at ground potential even if the insulation of the ungrounded, or hot, wire fails. This keeps personnel safe from shock when it is connected properly. The PLC in Figure 13.5(b) shows an unsafe situation, where the equipment ground conductor is opened or not connected, thus letting the chassis become hot when the insulation fails. The person touching the chassis will be shocked, and it could possibly be lethal under the right circumstances. Because you rely on a good ground for safety, make sure of the true grounding point by checking the ground point for low resistance.

When connecting cables to ground, make sure conducting surfaces to be connected are clean of oxides, paint, oils, or any material that would prevent a good connection. Use star washers, which bite into the metals and thus result in a good connection between ground cable and grounding point. When you have done the grounding properly with clean connections, you will not have to worry about interference.

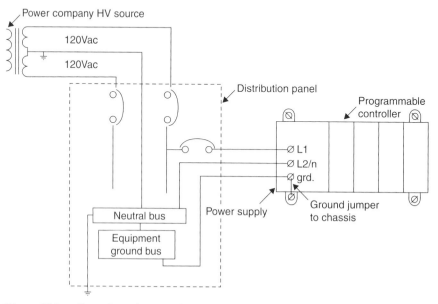

Figure 13.5a   Ground requirements for safety.

There are several ways interference is generated when low-voltage and low-current signals are present. Low means in the millivolt and milliampere range, and these types of signals are commonly found going to and from transducers. Interference may occur through capacitive coupling, for example, when power-source conductors are run in the same wire bundle as the conductors carrying low-level signals. It may also occur when something introduces a changing magnetic field that goes through the loop carrying low-level signals. Both these problems can be stopped by surrounding the wire carrying the low-level signal with a grounded shield. See Figure 13.6.

You must be careful not to make ground loops, which will cause nearby magnetic fields to introduce interference in the shield that capacitively couple to the signal wire. The way to stop ground-loop interference is to ground the shield at one end only. Remember, if you ground the shield at both ends, you create a loop, which could result in interference.

Figure 13.7 shows a good way to connect shields to prevent loops that also provides an easy way to detect if a shield is grounded more than once. The ground bus is insulated from ground by using insulating bushings and then grounded only once. It is grounded this way so you can check for extraneous grounds. If you remove the single ground and then check from the bus to ground, if there are any grounds at any point along any of the coaxes, your ohmmeter will read low. If there are no extra grounds, the ohmmeter will read infinity or very high. You can easily find where the coax is grounded by lifting the grounds one at a time until the ohmmeter reads high.

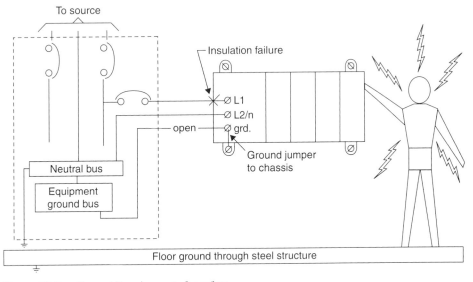

Figure 13.5b Ground Requirements for safety.

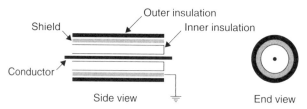

Figure 13.6 Coax cable.

Because there is potential for low-voltage signals to pick up interference from surrounding power cables, it is a good idea to separate low-voltage signals and power cables.

When handling modules, processors, and electronic plug-ins, you must be careful not to damage the internal electronics by discharging static electricity. You can get wrist ground straps to prevent this from computer suppliers.

When inserting modules, always turn off the power to the chassis so that a misalignment of a connector does not cause a short or result in an improper voltage being applied to the electronics.

Never use force when inserting or removing connectors, modules, or plug-ins. Often they are polarized or keyed so that the wrong connections cannot be made. Sometimes plastic catches are used to hold parts in place, and these can be easily damaged by using force.

You need to decide how to handle power outages. If you have a power failure and do nothing, the PLC will usually restart when the power returns. You can usually inhibit startup after a power failure by setting a switch in the chassis or a bit in the status file. This causes a major fault to be set automatically after a power outage so that the processor shuts down. If you want to restart after you set an inhibit bit, you must write a fault routine to clear the major fault bit. If you have a fault routine, you must tell the processor where the fault routine is located. You also have to tell the processor where to start in the program that was originally running—either the beginning or where you left off. There usually are safety issues that also need to be addressed on startup after power outages.

One safety issue is emergency shutdown when there is a need to protect personnel from being injured or equipment from being damaged. PLCs have an excellent record of reliable operation, but that does not mean they will never fail. An emergency could be caused by a module failure, by an improperly designed control program, or by the PLC itself operating improperly.

One way to handle an emergency is to cut the power to the modules controlling the process. The relay control circuit in Figure 13.8 is designed to control the power going to and from the I/O modules. A start push button causes the power to be applied to the modules through the closing of the normally opened master control relay contacts on rung 4. The emergency-stop push buttons are strategically

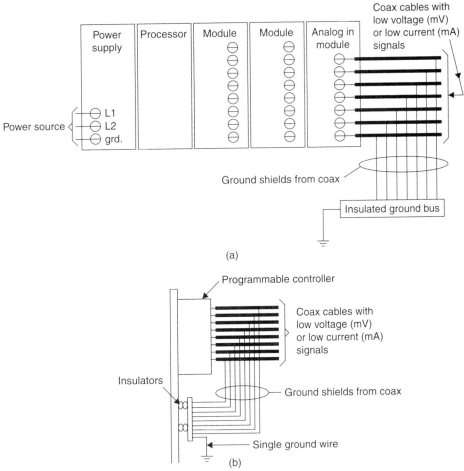

Figure 13.7 (a) Front view; (b) Side view.

placed where required throughout the control process. Two emergency stops are shown in the control, but you can put in as many as you need wherever you need them so if someone pushes an emergency stop, the master control relay will be deenergized and interrupt the power to the I/O modules. Output modules need a power source and input modules need a neutral, or common, return. This control is shown to give an example of a possible emergency stop. Every control situation needs to be evaluated for safe emergency shutdown. Your process may require a different control circuit.

Make sure the power source is not subject to resistive interference. Resistive interference occurs in a home with an inadequate power source when a freezer (or something that draws a large amount of current) causes the lights to dim. The voltage drop across the internal impedance of the source is too high and causes the

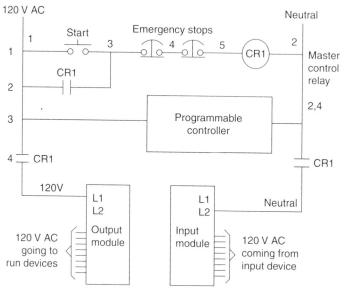

Figure 13.8 Emergency shutdown control.

voltage available at its terminals to decrease too much. You do not want your programmable controller to have a similar problem. You should have a clean power source for the programmable controller. This means devices on the same circuit should not degrade the voltage beyond a point where the processor sets a major fault and starts going into a power-outage mode.

External connections have to be made to the modules, and then these connections need to be checked out before the process is put in the run mode. This can be done by putting the PLC in the test mode. The control program gives an indication of the status of inputs and shows what could happen to the outputs, but it does not physically turn on the outputs. This mode enables you to operate input devices and to see if the indication you expect happens. This enables you to see if your addressing is correct.

▶ 13.3 STARTUP AND TROUBLESHOOTING

Initial-run tests require that you think about the safety of personnel and equipment. Make sure you have a means of disabling the output, such as Figure 13.8, if something goes wrong. Keep nonessential personnel clear of the process until you are done testing and know the control is working properly.

There is no substitute for good documentation, and it is critical in troubleshooting complex control. Your program should be structured and not done by trial and error. You should document the purpose of rungs by using the rung remarks. You will find a state diagram to be very useful. You should make tables to show the use of internal and external inputs and outputs. A table indicating the purpose of special words or bits used in your program is very helpful.

Check to be certain that input and output terminal labels match the labels used in the documentation of the control. Using different designations causes confusion and can result in connection mistakes.

Use a voltmeter to make sure the field terminals of the PLC and modules have the correct power applied to the source terminal. An output module usually requires an external source, and input modules require a common return. Without these connections the I/O modules cannot function. Check that the modules are in the right slot per the addressing in the control scheme. Plugging a TTL module into a slot wire for a 120-V output could result in a TTL module receiving 120 V and damaging the electronics in the module.

It is easy for incorrect addressing of inputs and outputs to occur. The longer the program is, the more likely this will happen. If an output shows that it is energized on the monitor screen and you have checked that the module is powered correctly, but no light is on at the output module indicators, then it is likely that the address is wrong. You can use a forcing function to check outputs, but you must be aware that a forced output may operate not under the control of the program.

Another common problem is that the program may skip over some control programming without your realizing it. This can occur when you use structured programming or a processor that allows more than one main program. If you have multiple main programs, the configure instructions for this feature tell which programs are going to run. You may not have configured the program you are troubleshooting. Structured programming is set up to skip over zones or subroutines to save on scan time and to execute only the states that need to be active. A mistake in transferring from one subroutine to another will also cause this type of problem.

When watching a monitor to see your program operate in real time, you need to realize that the program can execute faster than the screen can be updated. Because of this, you will not see an event such as an output being picked up for a few milliseconds. One way to find out if you are executing an area of control is to insert an unconditional retentive output in the area in question. The output will be set if you scanned that area.

Duplicating nonretentive outputs will cause problems and is not a good idea because the rungs controlling the duplicated nonretentive outputs may be telling one output to be on via one rung and off via another. The PLC will update the output as determined by the last rung executed. If you duplicate any nonretentive outputs in more than one zone when using zone control and only one zone is active, you may not get the response you want. This happens because the inactive zone turns the output off even though the active zone may be trying to turn it on. When a zone is inactive, all rungs within that zone are considered false by the processor.

If you have timers that are not functioning, make sure you have not addressed two timers with the same address. If one timer instruction sees a false rung and the other timer instruction sees a true rung, the false rung will constantly reset the duplicated timer, resulting in the timer instruction with the true rung not being able to time.

Because counters are retentive, they cannot function properly if they are never reset. If a counter keeps counting up well beyond its preset, you may have forgotten the reset. On the other hand, counters can be stopped from counting if their reset is held true, so if you have a nonworking counter, check the reset.

If you find changes that need to be made during the course of troubleshooting, it is critical that you update your documentation. If you make changes without doing this, the next person to use the documentation will have incorrect information. This could be minor, but it could also lead to serious consequences, such as when an employee makes changes and then leaves the company.

▶ 13.4 MAINTENANCE

PLCs have an excellent history of reliable operation, which is part of the reason they have been so successful. The required maintenance is minimal. There are some things that should be checked periodically.

- Clean the PLC periodically to remove excessive dust and dirt.
- Inspect terminals, modules, and communication connectors for looseness or possible conductor breakage. Vibration and expansion or contraction due to thermocycling can cause these items to become loose.
- Replace batteries per the manufacturer's recommendations or whenever the processor loses memory after being shut down.
- Make sure vents are not covered by items placed on the PLC or its enclosure.
- Keep a set of spare parts, such as fuses, terminals, screws, terminal-marker strips, and spare modules, that are critical should one fail.
- Always update documentation when making changes. Make hard copies as well as computer-backup copies of your programs. You never know when something will cause you to lose a program.
- Check enclosures to be certain that they are accomplishing their intended function. Check to see that the doors of the enclosure are not inadvertently left open, filters for air flow are replaced as necessary, and hazardous enclosures are properly sealed.
- Check to make sure that the environment for the PLC has not changed in a way that would subject the PLC to an unfriendly environment.

SUMMARY

A good way to ensure you have considered the important details for selecting a PLC is to develop a checklist. This chapter gives you a starting point for doing this. The checklist in this chapter may be sufficient, or you may have some additional constraints or requirements. If you do your homework by going through the checklist, you will have a much better chance of making the right choice. You can use the checklist as a specification by sending it out with a quote letter and letting the checklist dictate what is required.

▶ EXERCISES

You will have to do some research to answer these questions. There is more than one possible solution to some of the questions.

1. Using the library, trade journals, magazines, or the Internet, name five manufacturers of PLCs.

2. Describe Device Net.

3. Find a manufacturer who makes an I/O analog module with two channels of analog in and two channels of analog out in a single module. The analog-in channels are to be 4 to 20 mA and analog-out channels are to be –10 V to +10 V dc.

4. What Allen-Bradley SLC 500–series processor or processors can communicate with the Ethernet?

5. What Allen-Bradley SLC 500–series processor has a built-in RS232 channel to support ASCII for connection to other ASCII devices such as bar-code readers, scales, or printers?

6. What Allen-Bradley SLC 500 processor has the PID instruction with 32-bit resolution?

7. What memory modules are available for the Allen-Bradley SLC 500–series processors?

8. What device on the Allen-Bradley SLC 500 fixed processor lets you increase the I/O range?

9. Can a program made with Allen-Bradley APS software for the SLC 500 be read by RS Logix 500 software and converted to RS Logix 500 format?

10. What Allen-Bradley SLC 500 processor is designed specifically for the plastics industry and contains ERC2 algorithms for plastic machinery control?

11. For what does the acronym SCADA stand and what does it have to do with programming?

12. Troubleshoot the control in Figure 13.9, which should cause output O2/0 to turn off for 5 s and on for 1/2 s.

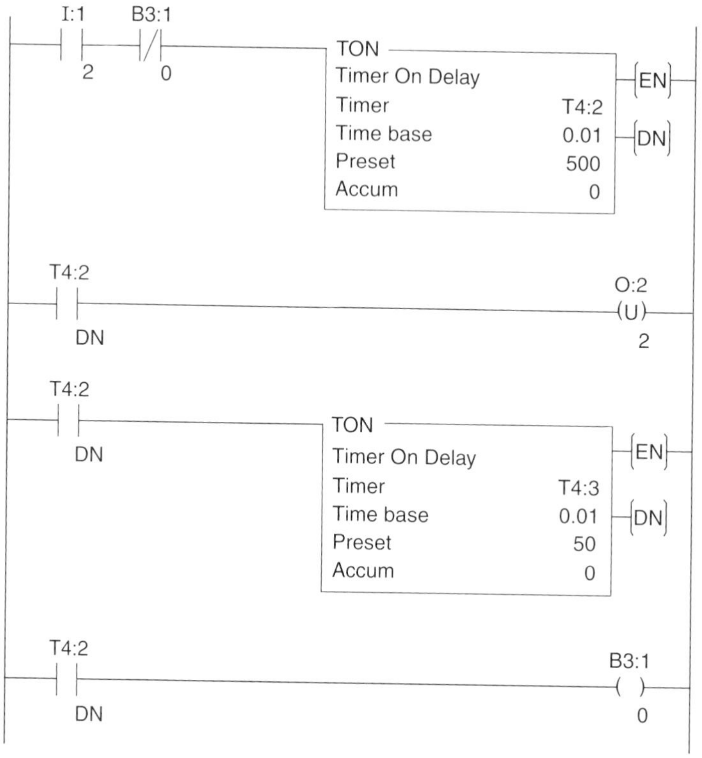

Figure 13.9 Exercise 12.

APPENDICES

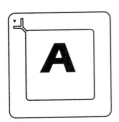

Relay Logic

INTRODUCTION

The programming language used by PLCs is based on the same symbols and functions used for control before PLCs were available. This language was specifically developed to make it easy for control people to change from relay control to computer control. This program language, called ladder logic, is popular because it is graphic and easy to troubleshoot. Appendix A looks at its roots by examining how relays can be used to make logical decisions.

257

▶ A.1 BASIC CONSTRUCTION OF A SIMPLE RELAY

The heart of control is the relay, which is a simple electromechanical device that has been around since the mid 1800s. One reason to study relay control is that if you replace existing relay control with a PLC, you would need to study the relay control to see how it works before you create the programming for a PLC.

Figure A.1 shows one form of design. Contact buttons are attached to a moving conducting bar, which could be copper, to make good contact. These buttons are usually made of copper and then silver- or gold-plated. Silver or gold plating works better than raw copper because the plating does not corrode easily as copper does. The conducting bar comes in contact with fixed contact buttons that are connected to screw terminals. The bar completes an electric circuit by touching these buttons. In Figure A.1, the coil is deenergized because the switch in series with the source is open. The bar is touching the bottom two fixed buttons, which are defined as normally closed because the coil of the relay is deenergized and they are closed. The upper set of contacts is normally open when the coil is deenergized. The state of the contacts when the relay coil is deenergized is defined as the normal state.

If you throw the switch, the coil is energized and the iron surrounded by the coil becomes a magnet, attracting the iron on the bar. This moves the bar upward, which causes the bar to break contact with the normally closed buttons and make contact with the normally open buttons. The feature that makes the relay so usable in control is that it is simple and it provides electrical isolation between the source used to pick up the coil and the source connected to the contacts. It usually takes just a few milliamps to energize the coil and the contacts can be used with a different source and can be made to handle large currents. For example, you might use a relay to start a 50-hp motor at 480 V ac. The motor draws a thousand times more current than required to pick up the coil of the relay. The person operating the switch is exposed to 120 V ac and currents of just a few milliamps but not to the 480 V ac. The isolation makes the control much safer for the operator.

Figure A.2 shows a different construction, in which the opening and closing of contacts is accomplished by a beam moving up and down. The spring pulls the left side

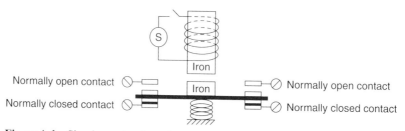

Figure A.1   Simple moving-bar relay.

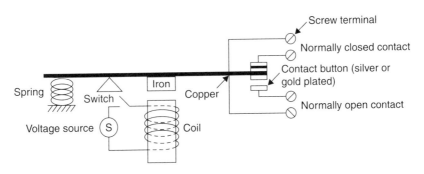

Figure A.2 Balance-beam relay.

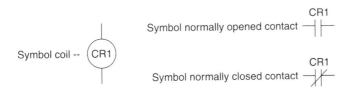

Figure A.3 Electric symbols for balance-beam relay.

of the beam down when the coil is deenergized, and the coil pulls the right side of the beam down when it is energized.

Figure A.3 shows the symbols used for the coil and the contacts in an electrical schematic. Notice that the contact symbols and the coil must have the same designation to tie them together.

▶ A.2 LATCH AND UNLATCH RELAY CONSTRUCTION

Figure A.4 shows a balance-beam latching relay. If a relay is picked up or a coil is energized, you may not want it to go back to its normal deenergized position. You may simply want to pulse it on and have it stay on. The relay in Figure A.4 causes a rod attached to a beam to push a key into a keyhole, which prevents the beam from returning to its deenergized position when the coil is turned off. You cannot pull the spring down its end of the beam unless the keyhole is pulled away. The keyhole is pulled away in this case by energizing a separate unlatch coil. There are two coils for this type of relay—one to latch the relay and one to unlatch the relay. Latching the relay makes it act as if it is permanently energized, even when it is not, which can act as a memory device to detect if some function has occurred even if it happens only briefly. The symbols for the latch and unlatch coils are shown for each of these devices in the figure.

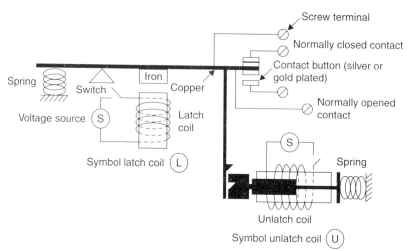

Figure A.4 Balance-beam latching relay.

► A.3 TIMER RELAY CONSTRUCTION AND OPERATION

Figures A.5 and A.6 show a timing device that can be attached to or piggybacked onto a relay. This type of timing unit is hydraulically operated. A piston is enclosed in a cylinder that is full of oil, and the piston can move only if the oil in front of it can move also. The oil can move fast through a one-way valve if the oil is going in the direction that unseats the ball. The ball is held in its seat by a spring, and if the oil is pushing in the same direction as the spring, it seals off the valve. Oil moving in the opposite direction can move the ball against the spring and thus open the valve. This valve permits oil flow in only one direction.

In Figure A.5, the piston can move only down because of the direction of the one-way valve. The figure shows another valve that can be cranked open and closed manually and that can bypass the one-way valve. How fast the piston can travel up or down depends on the one-way valve and how far the bypass valve is opened. The one-way valve opens automatically in one direction and closes automatically in the other direction, so the piston moves down fast when the oil flow unseats the ball. Otherwise, the piston cannot move up at all unless there is a bypass valve. The speed of the piston moving up depends on how far you open the valve. Thus, you can control how long it takes the timing unit to move up.

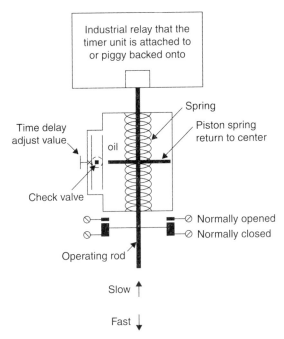

Figure A.5 Time delay on.

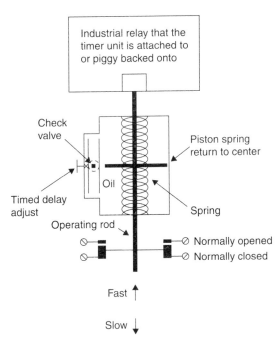

Figure A.6 Time delay off.

The wheel on the bypass valve is connected to a gauge that shows the time delay in seconds, minutes, or hours. Assume the mechanism used to connect the piggyback relay operating rod to the time-delay unit operating rod is such that it will pull the timing unit rod through a spring connection. When the coil of the piggybacked relay is energized, it pulls the rod of the timing unit up, and if there is little resistance from the time unit, it pulls the rod with it quickly, without any delay. When the piggybacked coil is deenergized, the timing unit rod is free to travel back to its initial position. The direction of the one-way valve determines the direction of delay in following the piggybacked relay.

In Figure A.6, the only change from Figure A.5 is that the one-way valve has been turned around. This reversal causes the delay to be in deenergizing the relay coil rather than energizing the coil. Notice the fast and slow directions of the piston have reversed from Figure A.5.

This is an interesting timing device because it is not electric but hydraulic. It can be made to do precise timing, is impervious to electrical noise, and is excellent in hostile environments such as high or low temperatures or dusty, dirty locations. There are other types of timers, such as electronic and pneumatic timers. Each has its own niche with corresponding advantages versus disadvantages. They are all shown by the same symbols in ladder logic and all operate in the same way.

Figure A.7 shows the operation of the piggybacked relay contacts and the timing unit contacts. The piggybacked relay contacts are not delayed and are independent from the timing unit. The timing unit contacts are dependent on the piggybacked operations, and there are two different types of normally opened and normally closed contacts. One set of contacts delays changing states on energizing the piggybacked relay coil and the other set of contacts delays changing states on deenergizing the piggybacked relay coil. Keep in mind that the timing unit has no coil of its own.

► A.4 RELAY LADDER-LOGIC SYMBOLS

Figure A.8 shows some additional symbols used in relay control circuits. These were taken from the NMTBA (National Machine Tool Builders Association) and JIC (Joint International Congress) standards.

► A.5 JIC LADDER-LOGIC DOCUMENTATION

Figure A.9 is a typical relay ladder-logic schematic, which looks somewhat like a ladder. It has two vertical lines, which represent power buses. The control is shown

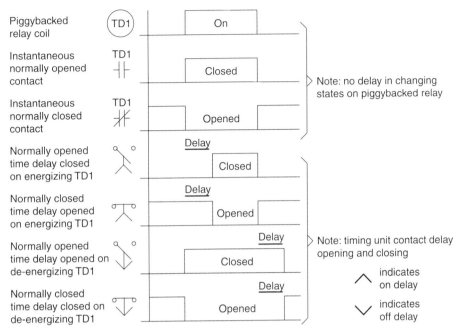

Figure A.7 Symbols and timing diagram.

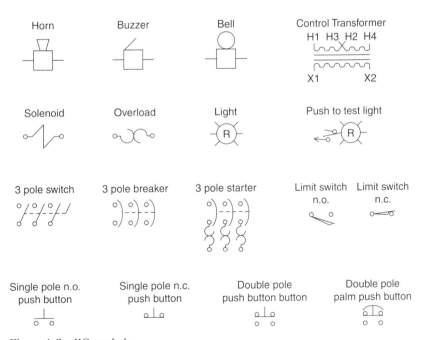

Figure A.8 JIC symbols.

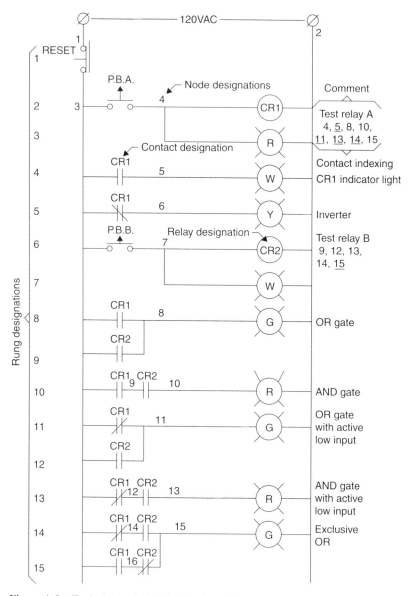

Figure A.9 Typical relay ladder-logic schematic.

264

horizontally between these buses. Each horizontal control line looks like a rung of a ladder. The control schematic is labeled per using *JIC* standards.

Rung Designation.   Each rung is given a number so it can be referred to when talking about the control. For example, relay CR1 has its coil on rung 2.

Node Designation.   Every node has a number designation. A node occurs where two or more devices are connected. Therefore, you can label wires so that when you are testing, you know at which node you are located.

Relay Designation.   This designation distinguishes one relay coil from another.

Comment.   Each rung can have a comment about its purpose. It is particularly important to make comments about each relay coil so that you know its intended purpose.

Contact Indexing.   Contact indexing shows where all the contacts for a specific relay are located and what type of contact is being used. You do this by giving the rung the contact is on. If it is underlined, it is a normally closed contact; otherwise it is a normally open contact.

Contact Designation.   The contact designation tells what relay the contact is associated with.

▶ A.6 RELAY LOGIC GATES

These control circuits are called relay logic because it is easy to make logic gates with relays. Consider the gates shown in the schematic of Figure A.9.

Rung 5 contains an inverter. The light will be off when A is pushed and on when A is not pushed.

Rungs 8 and 9 contain an OR gate. OR gates are made by placing contacts in parallel and then having the contacts pick up an output device. Note that if you press push button A *or* B *or* both, the green light will come on.

Rung 10 contains an AND gate. AND gates are made by placing contacts in series. You must push both A *and* B to turn on the light.

Rungs 11 and 12 contain an OR gate with an active low input or a not input. You must either not press push button A *or* press push button B for the light to come on.

Rung 13 contains an AND gate with an active low input. You must not press A *and* must press B to get the light to come on.

Rungs 14 and 15 contain an Exclusive OR gate. You must press A or B but not both to get the light to come on.

These gates are easy to make and see in the control schematic once you have the experience. You already have all the tools to make a computer, but it would take 10,000,000 gates to make a modern computer. To make that many gates would take 100,000,000 relays, which would cost a lot, take up two or three floors of a building, and use huge amounts of power. However, each relay can be replaced by controlling 1 bit in memory for the coil and 1 bit for each contact because 0 can represent off and 1, on. The memory chips available today make it inexpensive and easy to construct a modern computer.

▶ A.7 EXAMPLE RELAY CONTROL SCHEME

Figure A.10 shows a control scheme that operates a motor. Notice there are two distinct areas in this control schematic. The top is a power three-line circuit containing a circuit breaker, contacts from a motor starter, overloads, and the terminals for a motor. All these devices are in series and are fed by a three-phase source. This circuit is fed by voltages that are dangerous and possibly lethal for humans physically to contact, and the current may also be very high. The isolation transformer and the motor *starter* isolate the person operating the control from the high voltage and let the control portion be operated at 120 V ac. All the rungs below the isolation transformer are considered control. A PLC can replace the control part of the circuit but not the power three-line equipment.

This scheme illustrates start–stop Jog control. You do not need a PLC for this type of control because it probably will not change and is simple and inexpensive. Keep in mind that the PLC cannot replace the starter because the contacts must carry high current and must connect high-voltage sources to motors, etc.

Finally, Figure A.11 shows a relay control scheme involving a timing circuit that makes a light blink off and on using two timing relays. A timing diagram is also given to show how the circuit works. This programming can be done very easily on a PLC; a small PLC would be less expensive and would still be adequate.

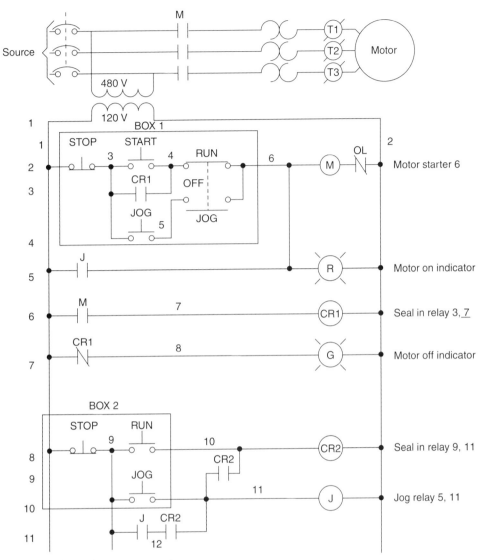

Figure A.10 Relay motor control.

SUMMARY

There are two different categories of relays: control and power. The PLC can replace a control relay and create huge savings. The ease of changing a PLC program versus having to rewire relay control makes relay control virtually obsolete. There is still great demand for power relays that can handle high currents and

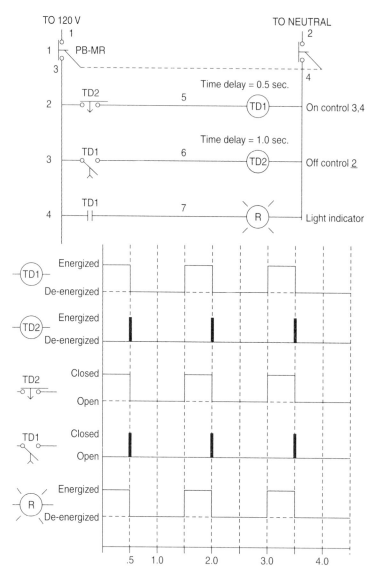

Figure A.11 Timing diagram.

voltages and give good isolation between the power circuit and control. PLCs cannot replace *contactors* and starters because of the current and voltage ratings required for these devices. Modern modules still use relays for outputs, which give a dry contact—dry because it is an electrically isolated contact to which you can apply whatever source you wish within the ratings of the contact. This is convenient because one module can be used for many different sources.

SLC 500 Series Status File

This appendix provides an overview of the status file that lets the user peruse the status file categories and functions on a word and bit basis. This information comes from Allen-Bradley's reference manual publication 1747-6.15 SLC 500 and Micrologix 1000 Instruction Set. It does not give the details you will find in the manual, but it does let you see in general what is in the status file. If you need specific details on how your processor uses these bits, you need to go to the instruction manual. Some bits may not be used or are handled differently, depending on which processor you have.

Word	Bit	Description
S:0		Arithmetic flags
	0	Carry bit
	1	Overflow bit
	2	Zero bit
	3	Sign bit
	4–15	Reserved
S:1		Processor mode/status/control
	0–4	4-bit function code telling what the processor is presently doing. Examples are download or run mode.
	5	Force enable bit
	6	Force installed bit
	7	Communication active bit
	8	Fault override at power-up bit
	9	Startup protection fault bit
	10	Load memory module on memory error bit
	11	Load memory module always bit
	12	Load memory module and run bit
	13	Major error halted bit
	14	Access denied bit
	15	First-pass bit

Word	Bit	Description
S:2		Selectable timed interrupt/DH485 communication
	0	STI pending bit
	1	STI enable bit
	2	STI execute bit
	3	Index addressing file range bit
	4	Saved with single-step test-enable bit
	5	Incoming command pending bit
	6	Message reply pending bit
	7	Outgoing message command pending bit
	8	Common interface file addressing mode
	9	Memory module program compare
	10	STI resolution selection bit
	11	Discrete input interrupt pending bit
	12	Discrete input interrupt enable bit
	13	Discrete input interrupt executing bit
	14	Math overflow selection bit
	15	Communication servicing selection bit
S:3L		Current/last scan time
	0–15	Low byte of latest scan time (resolution $\pm$ 10 ms)
S:3H		Watchdog scan time byte
	0–15	Low byte of latest scan time (resolution $\pm$ 10 ms)
	0–16	
S:4		Free-running clock
	0–16	Free-running clock time (resolution $\pm$ 10 ms)
S:5		Minor error code
	0	Overflow trap bit
	1	Reserved
	2	Control register error bit
	3	Major error detected while executing user fault routine
	4	M0–M1 referenced on disabled slot bit
	5–7	Reserved
	8	Memory module boot bit
	9	Memory module password mismatch bit
	10	STI overflow bit
	11	Battery-low bit
	12	Discrete input interrupt overflow bit
	13	Unsuccessful operating system load was attempted
	14	Channel 0 modem lost

Word	Bit	Description
	15	ASCII string manipulation
S:6		Major error code
	0–15	4-digit hexadecimal error code
S:7		Suspend code
	0–15	Suspend code
S:8		Suspend file
	0–15	Suspend program file number
S:9–10		Active nodes
	0–31	A bit file of the 32 possible active nodes
S:11–12		I/O slots enables
	0–31	A bit file of the 30 possible active slots
S:13–14		Math registers
	0–31	32-bit signed number register or 5-digit BCD
S:15L		Node address
	0–15	Node address
S:15H		Baud rate
	0–15	Baud rate
S:16–17		Test single-step start-step on-rung file
S:16	0–15	Executable rung number
S:17	0–15	Program file number
S:18–19		Test single-step breakpoint-rung file
S:18	0–15	Executable rung number
S:19	0–15	Program file number
S:20–21		Test-fault/power-down rung file
S:20	0–15	Executable rung number
S:21	0–15	Program file number
S:22		Maximum observed scan time
	0–15	Maximum observed scan time (resolution $\pm$ 10 ms)
S:23		Average scan time
	0–15	Average scan time (resolution $\pm$ 10 ms)
S:24		Index register
	0–15	Index register element offset value
S:25–26		I/O interrupt pending
	0–31	A bit file of the 30 possible active slots
S:27–28		I/O interrupt enabled
	0–31	A bit file of the 30 possible active slots
S:29		User fault routine number
	0–15	Program file number of user fault routine
S:30		Selectable time interrupt set point
	0–15	Time base value (resolution $\pm$ 10 ms)

Word	Bit	Description
S:30		Selectable time interrupt set point
	0–15	Time base value (resolution ± 10 ms)
S:31		Selectable time interrupt file number
	0–15	Selectable time interrupt program file number of user subroutine
S:32		I/O interrupt executing
	0–15	Slot number of the specialty I/O module that is generating the selectable time interrupt
S:33		Extended processor status and control word
	0	Incoming command pending, channel (ch.) 0
	1	Message reply pending, ch. 0
	2	Outgoing message command pending, ch. 0
	3	Selection status, ch. 0
	4	Communications active, ch. 0
	5	Communications servicing selection, ch. 0
	6	Message servicing selection, ch. 0
	7	Message servicing selection, ch. 1
	8	Interrupt latency control bit
	9	Scan toggle bit
	10	Discrete input interrupt reconfiguration bit
	11–12	2-bit code for on-line edit states
	13	Scan time base selection
	14	Channel 0 DTR control bit
	15	Channel 0 DTR force bit
S:34		Pass-through disable (504 only)
	0	DH+ to DH-485 pass-through disable bit
	1	DH+ active node table-enable bit
	2	Floating-point math-flag disable bit
	3	Global status word-transmit enable bit
	4	Global status word-receive enable bit
	5	DFI to DH+ pass-through enable bit
S:35		Last scan time (ms)
	0–15	Scan time (ms)
S:36		Extended minor error bits
	0–7	Reserved
	8	DII lost
	9	STI lost
	10	Memory modules data file overwrite Protection
	11–15	Reserved

Word	Bit	Description
S:37		Clock/calendar year
	0–15	Year value 0–65,535
S:38		Clock/calendar month
	0–15	Month value 1–12
S:39		Clock/calendar day
	0–15	Day value 1–31
S:40		Clock/calendar hours
	0–15	Hour value 0–23
S:41		Clock/calendar minutes
	0–15	Minutes value 0–59
S:42		Clock/calendar seconds
	0–15	Seconds value 0–59
S:43		Selectable timed interrupt timer
	0-15	Free-running timer with 10-μs increments
S:44		I/O event-interrupt timer
	0–15	Free-running timer with 10-μs increments
S:45		Discrete interrupt timer
	0–15	Free-running timer with 10-μs increments
S:46		Discrete input interrupt file number
	0–15	Program file number of DII subroutine
S:47		Discrete input interrupt slot number
	0–15	Slot number of the discrete I/O module that is generating the DII
S:48		Discrete input interrupt bit mask
	0–7	DII bit mask value
S:49		Discrete input interrupt compare value
	0–7	DII bit compare value
S:50		Discrete input interrupt preset value
	0–15	DII bit preset value
S:51		Discrete input interrupt return mask
	0–15	DII bit mask value
S:52		Discrete input interrupt accumulator value
	0–15	DII bit accumulator value
S:53–54		Reserved
S:55		Last discrete input interrupt scan time
	0–15	Last DII scan time (ms)
S:56 time		Maximum observed discrete input interrupt scan
	0–15	Maximum observed DII scan time (ms)
S:57		Operating system catalog number

Word	Bit	Description
	0–15	Operating system number
S:58		Operating system series number
	0–15	Operating system series number
S:59		Operating system FRN
	0–15	Operating system firmware release number
S:60		Processor catalog number
	0–15	Processor catalog number
S:61		Processor series
	0–15	Processor series number
S:62		Processor revision
	0–15	Processor revision number
S:63		User-type program
	0–15	Programming device number
S:64		User program functionality index
	0–15	Functionality level number
S:65		User RAM size
	0–15	RAM size in kilowords
S:66		User EEPROM size
	0–15	EEPROM size in kilowords
S:67–68		Ch. 0 active nodes
	0–31	Bit file of the 32 possible active nodes
S:69–82		DF1 half-duplex active node ch. 1
	0–223	Bit file of the 224 possible active nodes
S:83–86		DH+ active node ch. 1
	0–63	Bit file of the 64 possible active nodes
S:87–96		Reserved
S:97–98		Reserved
S:99		Global status word
	0–15	Global status value
S:100–163		Global status file
	0–1023	File of 64 words containing the global status word of all other nodes

C ControlLogix Status File

Unlike the SLC 500, the ControlLogix controller does not have a status file. The controller stores system data in objects. Two instructions are used to get system data and to set system data, the Get System Value (GSV) and Set System Value (SSV), both output instructions.

▶ C.1 GET SYSTEM VALUE (GSV) INSTRUCTION

The GSV instruction is an output instruction that retrieves data stored in an object and copies it into the destination.

Figure C.1 shows the GSV instruction. The operands that need to be entered are shown in Table C.1.

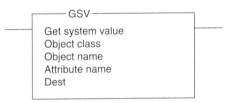

Figure C.1   GSV instruction.

TABLE C.1   GSV OPERANDS

Operand	Type	Format	Description
Object class		Name	Name of the object class
Object name		Name	Name of the specific object
Attribute name		Name	Attribute of the object
Destination	SINT INT DINT REAL	Tag	Destination of attribute data

► C.2 SET SYSTEM VALUE (SSV) INSTRUCTION

The SSV instruction sets controller system data that is stored in objects. The SSV instruction is shown in Figure C.2.

The operands that must be entered are shown in Table C.2.

The software lists the objects, object names, and attribute names for all the available attributes of the GSV instruction and the attributes that the SSV is allowed to set.

The following objects can be accessed:

AXIS	Status information from a servo module axis
CONTROLLER	Status information regarding a controller's execution
CONTROLLERDEVICE	Identification of the physical hardware of the controller
CST (coordinated system time)	Coordinated system time for the devices in a single chassis
DF1	An interface to the DF1 communication driver that can be configured for the serial port
FAULTLOG	Controller fault information

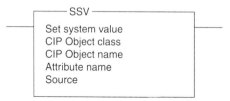

Figure C.2 SSV instruction.

TABLE C.2   SSV OPERANDS

Operand	Type	Format	Description
Object class		Name	Name of the object class
Object name		Name	Name of the specific object
Attribute name		Name	Attribute of the object
Source	SINT	Tag	Source tag that contains the data
	INT		to be copied to the attribute
	DINT		
	REAL		

MESSAGE	Tag name identifying the message; the interface to set up and trigger peer-to-peer communication
MODULE	Status information of a module
MOTIONGROUP	Status information about a group of axes for the servo module
PROGRAM	Status information about a program
ROUTINE	Status information about a routine
SERIALPORT	Interface to the serial communication port
TASK	Status information regarding a task
WALLCLOCKTIME	Time stamp used by a controller for scheduling

Each of these objects contains various attributes that may be monitored or set, depending on the attribute.

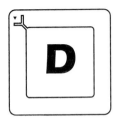

Answers to Odd-Numbered Problems

CHAPTER 1 PROGRAMMABLE CONTROLLERS

1. A typical desktop PC has a box containing the computer with a few slots inside for upgrading and for adding features by adding cards in slots. It also has a keyboard, disk drives, CD drive, modem, and monitor. A PLC looks different because there is no keyboard, disk drive, CD drive, modem, or monitor. Instead there is a box or chassis with units that go into slots. The power supply, processor, and other modules look pretty similar from the outside. The PLC has a key for making it run or not, and the modules have terminals to connect external devices to. The footprint of the PLC is usually much smaller than that of a PC.

3. Ladder logic is the most popular because it is a symbolic language developed specifically for the type of control that PLCs are designed to make easier. This graphical language can be watched on a monitor as it runs in real time and thus is easy to troubleshoot. Other languages have been introduced but have not enjoyed much success.

5. One reason for networking PLCs is so data can be collected at one source. This is accomplished by having a computer connected to the same network as the PLCs. Another reason for networking is to allow a PC or PLC to supervise the control of several PLCs at one time.

7. There are several reasons that a PLC is cost effective. One is that it is mass-produced using the latest microcomputer technology, and this technology is very competitive. Another is that the cost of making changes to a control process often involves reprogramming only the PLC. A third reason is that one inexpensive PLC can easily replace older relay control because the PLC does not have moving parts and hence does not wear out.

CHAPTER 2   INTERFACING AND LADDER-LOGIC FUNDAMENTALS

1.

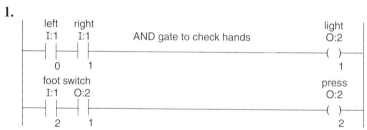

3.

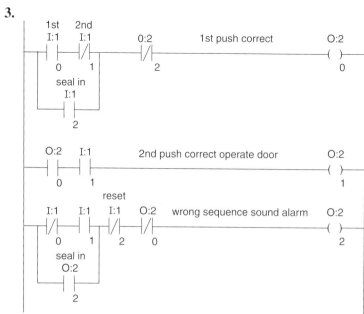

5. **a.** 1110111110110101—1110 1111 1011 0101—EFB5
   b. 1010110110110001—1010 1101 1011 0001—ADB1
   c. 1011110110110101—1011 1101 1011 0101—BDB5
   d. 1111110110111001—1111 1101 1011 1001—FDB9

7. Negative numbers are stored in 2's complement; therefore:

$$1111000011110000 \text{ (2's complement)}$$
$$0000111100001111 \text{ (1's complement)}$$
$$+1$$
$$0000111100010000 = -3856$$

9. The PLC executes the logic you have written rung by rung in order from top to bottom; because of this, you can control the order of events by the order of the rungs.

CHAPTER 3 SLC 500 PROCESSOR ARCHITECTURE VERSUS CONTROLLOGIX

1. 256 data files and 256 program files
3. Subroutines
5. I:3/05
7. T22:6.PRE
9. Integer: N
11. User assigned (B, T, C, R, N, F, ST, and A)
13. Continuous and periodic. There are 1 continuous and 32 periodic tasks; if there is no continuous task, there can be 32 periodic tasks, but if there is a continuous task, there can only be 31 periodic tasks.
15. Routine
17. Base, alias, and consumed
    REAL: 4-byte floating point
19. CONTROL: control structure for array instructions
    COUNTER: control structure for counter instructions
    MOTION_INSTRUCTION: control structure for motion instructions
    PID: control structure for the PID instruction
    TIMER: control structure for timer instructions
    AXIS: control structure for an axis (*controller tag only)

21.

Bit 31	24	23	16	15	87	0
Data allocation 1			Data_1			
Data allocation 2			Data_2			
Data allocation 3		Unused	Unused	Data_4	Data_3	
Data allocation 5		Unused	Unused		Bit 0 Bit_0	
					Bit 1 Bit_1	

23. Three
25. 120

CHAPTER 4 SLC 500 AND CONTROLLOGIX BIT INSTRUCTIONS

1. The symbol for a XIC instruction is —] [—.
   When there is a 1 in the data table at the bit address of the instruction, the instruction is logically true.
3. When the rung condition to the instruction is true, the instruction writes a 1 to the bit address of the instruction. When the rung condition is false, the instruction writes a 0.

5.

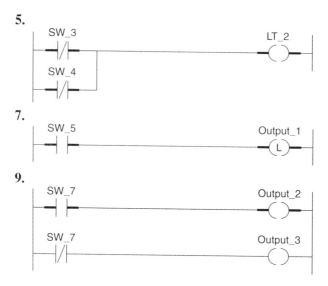

7.

9.

CHAPTER 5 SLC 500 AND CONTROLLOGIX TIMER AND COUNTER INSTRUCTIONS

1. TON timer. TOF timer resets when it goes true.
3. SLC 500: 32,767; ControlLogix: 2,147,483,647
5. SLC 500:

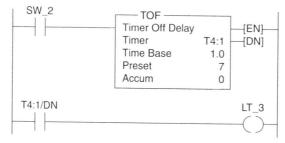

ControlLogix:

7. SLC 500:

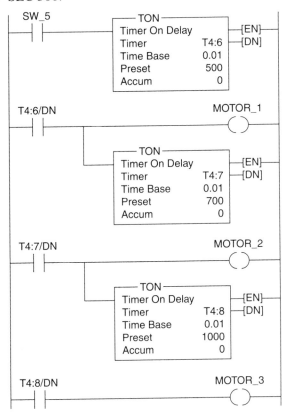

7. Continued

ControlLogix:

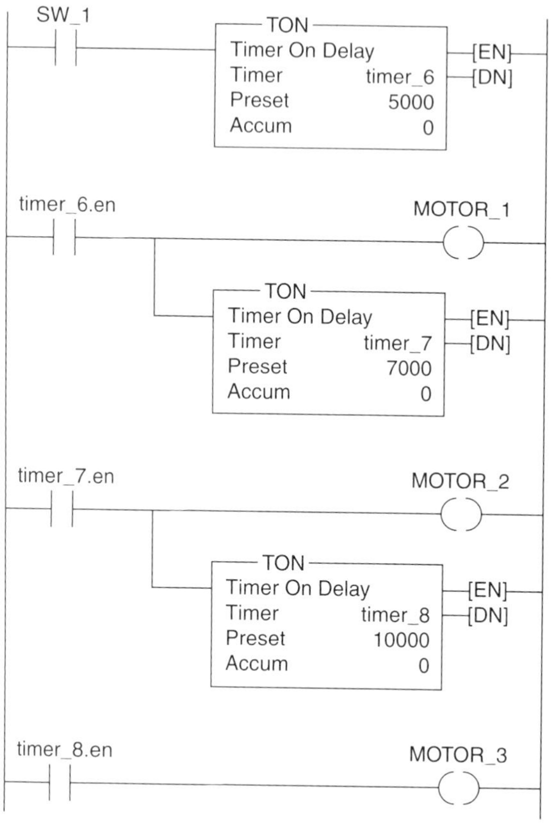

9. SLC 500:

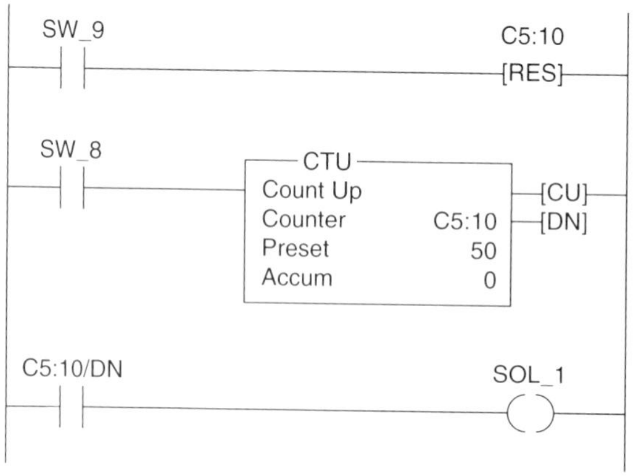

ControlLogix:

```
      sw_9                                    counter_1
      ┤ ├────────────────────────────────────[RE ]──────

      sw_8            ┌──── CTU ────┐
      ┤ ├─────────────│ Count Up    │────────────[CU]──
                      │ Counter  counter_1        [DN]
                      │ Preset      50  │
                      │ Accum        0  │
                      └─────────────────┘

   counter_1.dn                             SOL_1
      ┤ ├──────────────────────────────────────( )──────
```

CHAPTER 6   SLC 500 AND CONTROLLOGIX DATA COLLECTING

1. The MOV instruction copies the data in the source to the address in the destination. The source could be a constant or an address.

3. The mask in the MVM instruction determines the status of which bits in the source will be copied to the destination. A 1 in the mask will allow the status of the corresponding bit in the source to be copied to the destination. A 0 will block the copying of the corresponding bit, and the destination bit's status will remain in its last state.

5. The CLR instruction will zero the value at its address.

7. XOR

9.

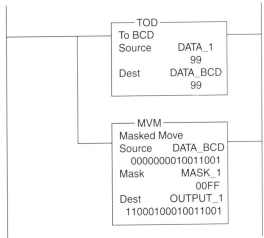

11.

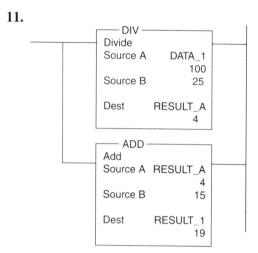

13.

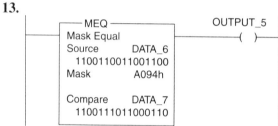

CHAPTER 7   SLC 500 AND CONTROLLOGIX
FILE-DATA MANIPULATION

1. All modes

3. #File type:file number: #N22:0

5.

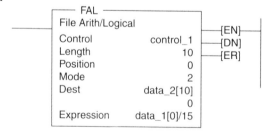

7. SLC 500:

```
  SW_1              ┌──COP─────────────────────┐
──┤ ├──────────────┤ Copy File                 │───
                    │ Source  #SETPOINT_IDLE    │
                    │ Dest              #OVEN   │
                    │ Length                5   │
                    └───────────────────────────┘

  SW_1              ┌──COP─────────────────────┐
──┤/├──────────────┤ Copy File                 │───
                    │ Source  #SETPOINT_RUN     │
                    │ Dest              #OVEN   │
                    │ Length                5   │
                    └───────────────────────────┘
```

ControlLogix:

```
  SW_1              ┌──COP─────────────────────┐
──┤ ├──────────────┤ Copy File                 │───
                    │ Source    setpoint_idle[0]│
                    │ Dest           oven[0]    │
                    │ Length              5     │
                    └───────────────────────────┘

  SW_1              ┌──COP─────────────────────┐
──┤/├──────────────┤ Copy File                 │───
                    │ Source    setpoint_run[0] │
                    │ Dest           oven[0]    │
                    │ Length              5     │
                    └───────────────────────────┘
```

CHAPTER 8   SLC 500 AND CONTROLLOGIX SHIFT REGISTER AND SEQUENCER INSTRUCTIONS

1. BSL or BSR

3. FFL and FFU

5.

```
  SW_6              ┌─BSR──────────────────┐
──┤ ├──────────────┤ Bit Shift Right       ├──(EN)──
                    │ File         #B22:0   ├──(DN)──
                    │ Control        R6:7   │
                    │ Bit address    SW_7   │
                    │ Length           10   │
                    └───────────────────────┘
```

7. SLC 500:

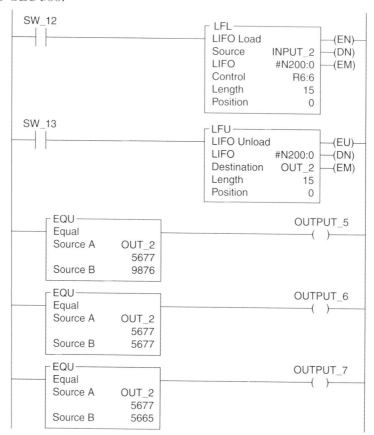

ControlLogix:

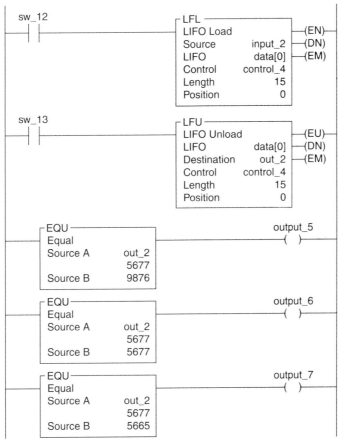

9.

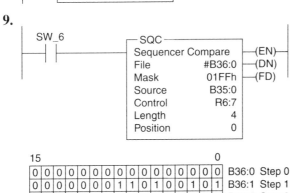

CHAPTER 9   SLC 500 AND CONTROLLOGIX
PROGRAM-CONTROL INSTRUCTIONS

1. A start rung with input conditions and an MCR instruction as the output plus an end rung with no input conditions and an MCR instruction as the output are required. The rungs between these two rungs are inside of the MCR zone.

2. To turn off all nonretentive instructions within the zone when the start rung goes false.

3. All nonretentive instructions go false.

4. All rungs within the zone are scanned and the logic is executed.

5. When SW_5 is turned off, the status of the outputs is as shown.

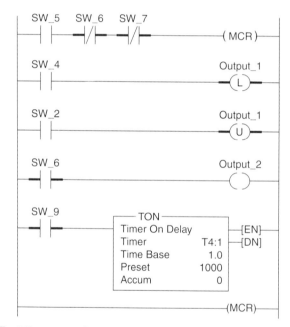

7. When a section of logic is jumped over and not scanned, all outputs in the section jumped over will remain in their last state.

9. The status of the outputs are the result of SW_8 being turned off. SW_4 is turned off, SW_6 is turned off, SW_9 is turned off, and SW_7 is turned off in that order.

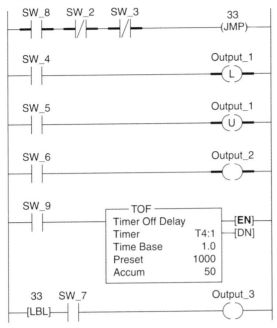

11. A nested subroutine is a subroutine that is jumped to from another subroutine.

13. The passing of parameters allow data to be copied between memory locations when jumping to a subroutine or when returning back to the point from which a jump to subroutine was executed.

CHAPTER 10   STRUCTURED PROGRAMMING ON THE PLC

1.

```
    O:2
 ───┤ ├──────────────────────────────────────(MCR)──   Start fence
     4
    I:1    O:2
 ───┤ ├────┤ ├─────────────────────────────┬──(L)───    Exit to
     2      5                               │   5        etch
                                            │  O:2
                                            └──(U)───    Turn off
                                               4         present state
    I:1    I:1                                  O:2
 ───┤ ├────┤ ├─────────────────────────────┬──(L)───    Exit to
     3      5                               │   6        clean
                                            │  O:2
                                            └──(U)───    Turn off
                                               4         present state
    I:1    I:1                                  O:2
 ───┤ ├────┤ ├─────────────────────────────┬──(L)───    Exit to
     4      5                               │   7        rinse
                                            │  O:2
                                            └──(U)───    Turn off
                                               4         present state
                                               O2:
 ──────────────────────────────────────────── (L)───    Lower
                                               13         motor
    I:1                                         O2:
 ───┤ ├───────────────────────────────────────(L)───    Emergency
     1                                          14        stop detection

 ──────────────────────────────────────────── (MCR)──   End fence
```

3.

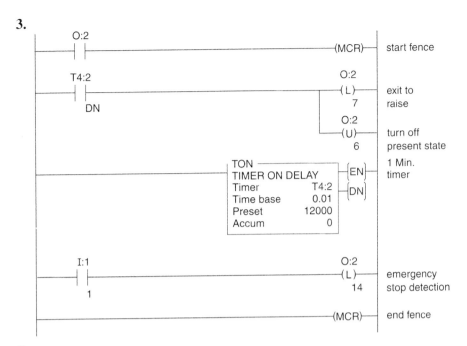

O:2 ─┤ ├─ (MCR)	start fence
T4:2 ─┤ ├─ DN → O:2 (L) 7	exit to raise
O:2 (U) 6	turn off present state
TON TIMER ON DELAY Timer T4:2 Time base 0.01 Preset 12000 Accum 0 (EN)(DN)	1 Min. timer
I:1 ─┤ ├─ 1 → O:2 (L) 14	emergency stop detection
(MCR)	end fence

5.

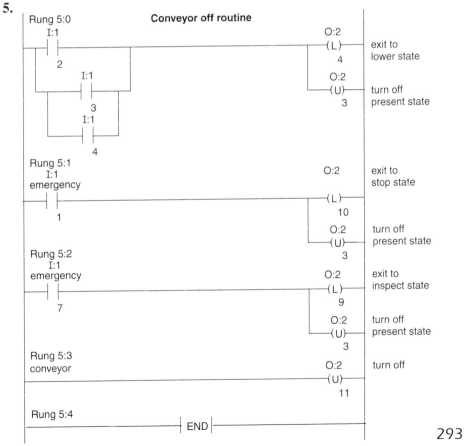

Conveyor off routine

Rung 5:0
I:1 ─┤ ├─ 2 / I:1 ─┤ ├─ 3 / I:1 ─┤ ├─ 4 → O:2 (L) 4 — exit to lower state
O:2 (U) 3 — turn off present state

Rung 5:1
I:1 emergency ─┤ ├─ 1 → O:2 (L) 10 — exit to stop state
O:2 (U) 3 — turn off present state

Rung 5:2
I:1 emergency ─┤ ├─ 7 → O:2 (L) 9 — exit to inspect state
O:2 (U) 3 — turn off present state

Rung 5:3
conveyor → O:2 (U) 11 — turn off

Rung 5:4
─┤ END ├─

293

7.

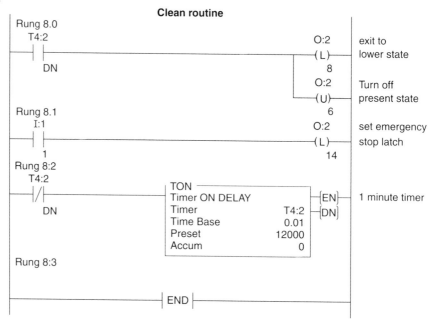

9.

Raise routine

Rung 10:0
```
  I:1    O:2                                    O:2      exit to
 ─┤ ├───┤ ├──────────────────────────────────(L)──    emergency stop
   6     14                                     10      state
                                               O:2      turn off
                                             ─(U)──     present state
                                               8
```

Rung 10:1
```
  I:1    O:2                                    O:2      exit to
 ─┤ ├───┤/├──────────────────────────────────(L)──    conveyor on
   6     14                                             state
                                               O:2      turn off
                                             ─(U)──     present state
                                               8
```

Rung 10:2
```
  I:1                                           O:2
 ─┤ ├─────────────────────────────────────────(L)──   set emergency
   1                                            14      stop latch
```

Rung 10:3
```
  O:2                                           O:2
 ─┤ ├─────────────────────────────────────────(L)──   raise moter
   8                                            12
```

Rung 10:4
```
 ──────────────────────────┤ END ├────────────────
```

11.

Rung 12:0

Emergency stop

```
| I:0                                          O:2    exit to
|─┤ ├──────────────────────────────────┐     ─(L)─   conveyor on
|   0                                    │      2     state
|                                        │
|                                        │     O:2    turn off
|                                        ├──── ─(U)─   present state
|                                        │      10
|                                        │
|                                        │     O:2    reset
|                                        └──── ─(U)─   stop latch
|                                                14
```

Rung 12:2

```
|──────────────────────┤ END ├─────────────────────────────|
```

13. It scans through the chart from left to right and from top to bottom.

15.

```
| Rung 13:0      File 13 Acknowledge Push Button
|  I:000
|──┤ ├─────────────────────────────────────────(EOT)──|
|   00
```

17.

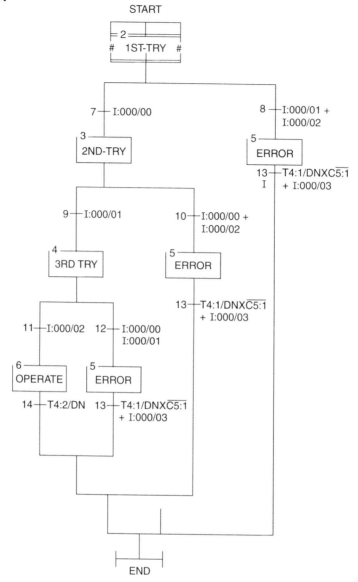

19.

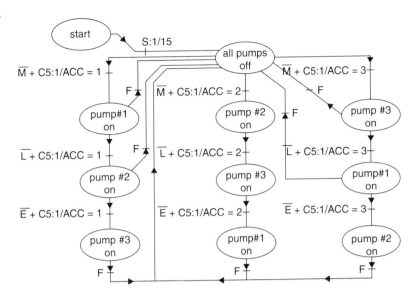

19. Continued

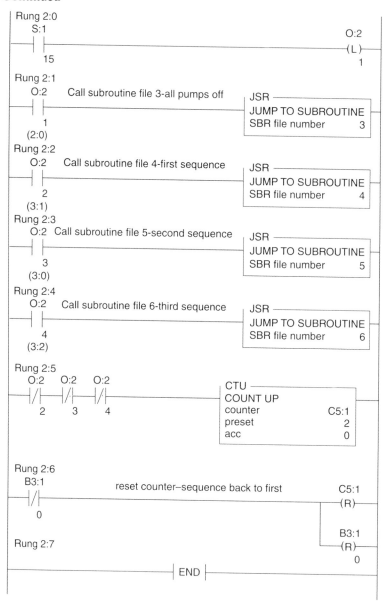

19. Continued

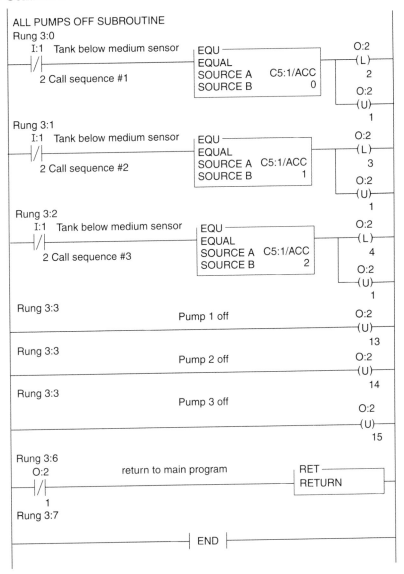

ALL PUMPS OFF SUBROUTINE
Rung 3:0

```
    I:1   Tank below medium sensor    EQU                        O:2
   ─┤/├──────────────────────────┐    EQUAL                      ─(L)─
         2 Call sequence #1       │    SOURCE A    C5:1/ACC         2
                                  │    SOURCE B           0       O:2
                                  │                               ─(U)─
                                                                    1
```

Rung 3:1

```
    I:1   Tank below medium sensor    EQU                        O:2
   ─┤/├──────────────────────────┐    EQUAL                      ─(L)─
         2 Call sequence #2       │    SOURCE A    C5:1/ACC         3
                                  │    SOURCE B           1       O:2
                                  │                               ─(U)─
                                                                    1
```

Rung 3:2

```
    I:1   Tank below medium sensor    EQU                        O:2
   ─┤/├──────────────────────────┐    EQUAL                      ─(L)─
         2 Call sequence #3       │    SOURCE A    C5:1/ACC         4
                                  │    SOURCE B           2       O:2
                                  │                               ─(U)─
                                                                    1
```

Rung 3:3

```
                          Pump 1 off                             O:2
   ──────────────────────────────────────────────────────────── ─(U)─
                                                                   13
```

Rung 3:3

```
                          Pump 2 off                             O:2
   ──────────────────────────────────────────────────────────── ─(U)─
                                                                   14
```

Rung 3:3

```
                          Pump 3 off
                                                                 O:2
   ──────────────────────────────────────────────────────────── ─(U)─
                                                                   15
```

Rung 3:6

```
    O:2               return to main program        RET
   ─┤/├───────────────────────────────────────────┐ RETURN
     1
```

Rung 3:7

```
   ────────────────────────────────┤ END ├────────
```

19. Continued

FIRST SEQUENCE SUBROUTINE

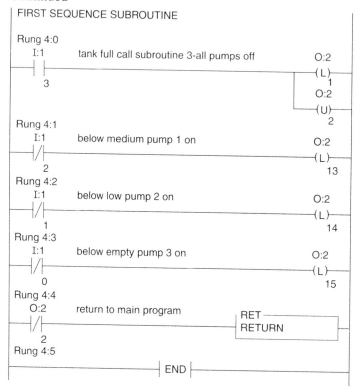

Rung 4:0

I:1   tank full call subroutine 3-all pumps off   O:2
3   —(L)—
                 1
                 O:2
                 —(U)—
                 2

Rung 4:1

I:1   below medium pump 1 on   O:2
2   —(L)—
                 13

Rung 4:2

I:1   below low pump 2 on   O:2
1   —(L)—
                 14

Rung 4:3

I:1   below empty pump 3 on   O:2
0   —(L)—
                 15

Rung 4:4

O:2   return to main program   RET
2   RETURN

Rung 4:5

—| END |—

19. Continued

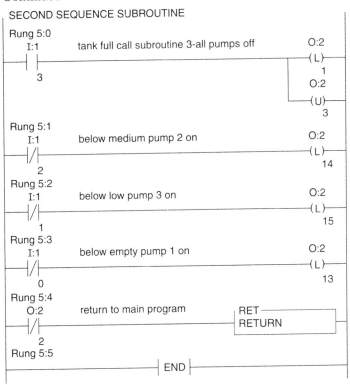

SECOND SEQUENCE SUBROUTINE

Rung 5:0

I:1   tank full call subroutine 3-all pumps off   O:2
—| |—   —(L)—
3   1
                                                      O:2
                                                   —(U)—
                                                   3

Rung 5:1

I:1   below medium pump 2 on   O:2
—|/|—   —(L)—
2   14

Rung 5:2

I:1   below low pump 3 on   O:2
—|/|—   —(L)—
1   15

Rung 5:3

I:1   below empty pump 1 on   O:2
—|/|—   —(L)—
0   13

Rung 5:4

O:2   return to main program   RET
—|/|—   RETURN
2

Rung 5:5

—| END |—

19. Continued

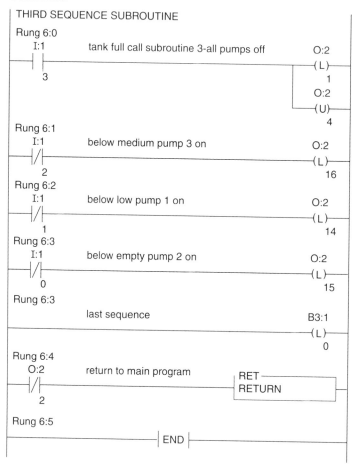

THIRD SEQUENCE SUBROUTINE

Rung 6:0
```
   I:1          tank full call subroutine 3-all pumps off      O:2
  ─┤ ├─                                                      ─(L)─
    3                                                           1
                                                              O:2
                                                            ─(U)─
                                                              4
```

Rung 6:1
```
   I:1                    below medium pump 3 on              O:2
  ─┤/├─                                                     ─(L)─
    2                                                         16
```

Rung 6:2
```
   I:1                    below low pump 1 on                 O:2
  ─┤/├─                                                     ─(L)─
    1                                                         14
```

Rung 6:3
```
   I:1                    below empty pump 2 on               O:2
  ─┤/├─                                                     ─(L)─
    0                                                         15
```

Rung 6:3
```
                         last sequence                      B3:1
                                                           ─(L)─
                                                             0
```

Rung 6:4
```
   O:2                    return to main program          ┌ RET ──────┐
  ─┤/├─                                                   │ RETURN     │
    2                                                     └────────────┘
```

Rung 6:5
```
                              ─┤ END ├─
```

CHAPTER 11 COMMUNICATION WITH OTHER PROGRAMMABLE CONTROLLERS AND COMPUTERS

1. 32 nodes numbered 0–31
3. 1747-AIC link couplers
5. Only the controller plugged into can be accessed.
7. SLC 5/04
9. 57.6k baud, 10,000 ft; 115.2k baud, 5000 ft; 230.4k baud, 2500 ft
11. SLC 5/05
13. There can be up to 99 addressable nodes on a link.

1.

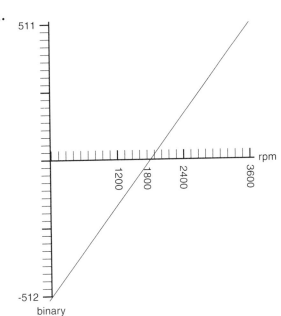

a. The maximum positive number is $01\ 1111\ 1111_2$ or 511_{10} and the maximum negative number in 2's complement form is $10\ 0000\ 0000_2$, or 512_{10}.

b.
$$\text{slope} = \frac{S_{max} - S_{min}}{\text{binary}_{max} - \text{binary}_{min}} = \frac{511 - (-512)}{3600\ \text{rpm} - 0\ \text{rpm}} = 0.2842/\text{rpm}$$

When $\quad x = 3600$ and $y = 511$, $b = y - mx = $ offset, or

offset $= 511 - 0.2842/\text{rpm} \times 3600\ \text{rpm} = -512$

binary $= mx + b = 0.2842/\text{rpm} \times 2500\ \text{rpm} - 512 = 199_{10} = 0011000111$

c. $11\ 0000\ 1111_2$ is a negative number stored in 2's complement form. Converting this to its decimal value requires taking it out of the 2's complements form. To do this, invert all the bits and add 1:

$11\ 0000\ 1111_2$

$00\ 1111\ 0000_2 + 1 = 00\ 1111\ 0001_2 = -241$

This value must be scaled binary to engineering units by referring to the following diagram. Notice the y-axis is where you want to go.

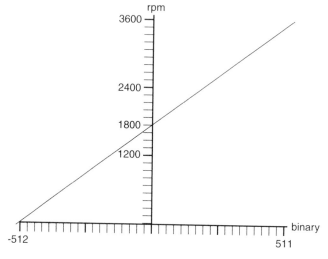

$$\text{slope} = \frac{S_{\max} - S_{\min}}{\text{binary}_{\max} - \text{binary}_{\min}} = \frac{3600 \text{ rpm} - 0 \text{ rpm}}{511 - (0 - 512)} = 3.519 \text{ rpm}$$

When $x = -512$ and $y = 0$, $b = y - mx = $ offset, or

offset $= 0$ rpm $- 3.519$ rpm $\times -512 = 1800$ rpm

$S = $ slope $\times$ binary $+$ offset $= 3.519$ rpm $\times (-241) + 1800$ rpm $= 952$ rpm

3.
$$\text{slope} = \frac{S_{\max} - S_{\min}}{\text{binary}_{\max} - \text{binary}_{\min}} = \frac{4095 - (-4096)}{32767 - (-32768)} = 0.125$$

PIDvalue $=$ slope $\times$ binary $+$ offset $= 0.125 \times (7575) + 0 = 946$

5.
$$G_p = \frac{\% \text{ change PV}}{\% \text{ change CV}} = \frac{30}{15} = 2$$

$$K_c = \frac{1}{K_p \tau_{\text{ratio}}} = \frac{1}{2 \times 4} = \frac{1}{8}$$

$$T_I = \Gamma_p = \frac{\text{dynamic settling time}}{4} = \frac{50,000 \text{ s}}{4} = 1250 \text{ s}$$

7.
$$K_C = 1.2/LS = 1.2(1.5 \text{ min} \times 0.5\text{psi/min}) = 1.6/\text{psi}$$

Note the units are per psi, but K_C is entered as a unitless number.

$$T_i = 2L \text{ (min)} = 2 \times 1.5 \text{ min} = 3 \text{ min}$$

$$T_d = 0.5L \text{ (min)} = 1.5 \text{ min}/2 = 0.75 \text{ min}$$

9. D, or differential, control is seldom used by itself because it looks at the rate of change of the error or process value and makes large corrections to fast changes. Large corrections can cause large overshoots, and this creates highly unstable control. Severe oscillation and equipment damage can easily result.

CHAPTER 13 PROGRAMMABLE CONTROLLERS: SELECTION, INSTALLATION, GROUNDING AND SAFETY, TROUBLESHOOTING, AND MAINTENANCE

1. ABB Industrial Systems Inc., Acroloop Motion Control Systems, Adatek, Advanced Micro Controls, Inc., Allen-Bradley, GE Fanuc, Fuji Electric, Klockner-Moeller, Mitsubishi Electric Automation Inc., Omron, Siemens, and many others.

3. Allen-Bradley high-resolution analog combination I/O cat #1746-NIO4V

5. The Allen-Bradley SLC 505 series processor has a built-in RS232 channel to support ACSII for connection to other ASCII devices such as bar-code readers, scales, or printers.

7. SLC 501 and 502-EEPROM and UVPROM, SLC 503, 504, and 505-EEPROM

9. Yes, a program made with Allen-Bradley APS software for the SLC 500 can be read by RS Logix 500 software and converted to RS Logix 500 format.

11. The acronym SCADA stands for supervisory control and data acquisition and is used to categorize the software and systems that many companies have for accomplishing this task. You could call up any automation company's sales office and ask what SCADA software or systems they have for PLCs. When you buy a PLC, it may have software to do SCADA. Allen-Bradley's software for doing this is RS view 32. It allows you to program a PLC to specifically do supervisory control and data acquisition.

.

Glossary

Address   A specific location in a computer's memory. In PLCs the address has the format of file.word/bit.

Alias tag   An alias tag references a memory location that has been defined by another tag.

Analog   A signal that is continuously varying instead of changing in steps when going between limits. Real-world signals are seen as changing continuously. An example would be a sinusoidal signal from an electric generator.

Analog input module   A module which conditions and isolates input signals so they are compatible with PLC's data bus. An example of conditioning would be changing 120Vac to a 5V digital signal.

Analog output module   A module which conditions and isolates PLC's signals so they are compatible with connected external devices. An example of conditioning would be changing a 5-V digital to a 48-V ac signal to run an external device.

AND gate   A logic device requiring all inputs to be satisfied before activating an output.

Antireset wind up   A large error can be produced by the integral term during startup and cause overshoot. The antireset windup refers to turning off the integral term of the PID function during start up.

Array   An array is a grouping of tags of similar data types in a Logix 5550 controller. A contiguous block in controller memory.

ASCII   An acronym for American Standard Code for Information Interchange. This is a 7-bit digital code for generating text.

ASCII input module   A module that converts text into 7-bit digital codes that are then in a form that the PLC can store and process.

ASCII output module   A module that converts alphanumeric information from PLC into 7-bit digital ASCII codes, which are then sent to external peripheral devices via one of the communication interfaces such as RS 232.

Automatic control   A process that uses feedback to maintain an output at a specific set point.

Barrel temperature module   A module that enables the user to monitor four zones of autotuned PID heat/cool for temperature control. Molding machines and extruders commonly use this module for controlling barrel temperature while injecting material. Make sure the processor you are using can support this module.

Base tag   A definition of the memory location where a data element is stored.

Baud rate   The number of elements per second. If the element is a bit, then the baud rate is given in bits/s.

BCD   An acronym for binary-coded decimal. It is a digital code for the decimal digits. Hundreds of codes have been developed. The most popular code is 8421.

BCD input module   A module that receives standard 8421 4-bit BCD, which the processor can recognize and store or process.

BCD output module   A module that sends standard 8421 4-bit BCD to external devices.

Binary   A number system that has only two digits, 0 and 1. It is easy to create two-state devices to represent 0 and 1. An example is a transistor; turned on it represents 1, and turned off it represents 0.

Bit   One binary digit, which must be 0 or 1.

BOOL   One bit boolean.

Branch   A parallel path in a program rung.

Chassis   The hardware enclosure that holds the modules and processor.

Consumed tag   A consumed tag references data that comes from another controller.

Contactor   A relay designed to handle currents above 10 A.

Controlled value   The output value that the automatic control adjusts to keep the process value at a set point. It is referred to as CV.

ControlNet   ControlNet is the primary communications network of Control-Logix. It combines the capabilities of DH+ and remote I/O networks. ControlNet is an open high-speed deterministic network that transfers time-critical I/O updates and controller to controller interlocking data and non–time-critical data such as data monitoring and program uploads and downloads on the same network.

Data files   An area of the processor's memory used to hold digital data that can be accessed and used by the user's programs.

Data Highway Plus (DH+)   An Allen-Bradley local information network that is designed for remote programming and for accessing and transferring data.

Deadband   A range of error in which the processor will not adjust the output via the PID instruction.

Device Net   A device used instead of hardwiring a device to the PC. You must have a device that has a microprocessor built into it to enable the device to communicate on the network. These devices are often referred to as smart devices.

DF-1 full duplex   Also referred to as DF1 point-to-point protocol. In full-duplex mode there can be simultaneous communication between two devices in two directions.

DF-1 half-duplex   A protocol that uses a master–slave-type communication. Communication takes place in one direction at a time.

DH-485   An Allen-Bradley proprietary token-passing network that supports a maximum of 32 devices with a maximum length of 4000 ft.

DINT   Four-byte integer.

Discrete control   Two-state control in which an input or output is either on or off.

Discrete input module   A module that receives two state inputs, which the processor can recognize and store or process.

Discrete output module   A module that receives discrete output information from the processor and uses this to send voltages to its terminals for operating devices attached.

EEPROM   An acronym for electrically erasable programmable read-only memory. This is read-only memory that is nonvolatile but that can be changed electrically with special programming.

Electromechanical Device   A device made up with mechanical and electrical components.

Element   A group of bits or words that are needed to define an instruction. An XIC requires 1 bit, but timers and counters require 3 words.

Error   The difference between the set point and the process value. There are two ways it is calculated, either the set point minus the process value or the process value minus the set point.

Ethernet   A protocol that determines the physical arrangement of connectors and the way communication will take place.

Exclusive OR gate   A two-input logic device requiring only one but not the other input to be satisfied before activating an output.

Gray code   A special code used mainly for position detection, in which only one bit changes as you move sequentially from one increment to the next.

Gray encoder module   A module that receives TTL gray-code signals from an input device and changes them to binary to be made available to the processor.

HEX Number   A number system with 16 digits. The digits are 0, 1, 2, 3, 4, 5, 6, 7, 8, 9, A, B, C, D, E, F.

High-speed counter encoder module   A module that enables you to count and encode faster than you can with a regular control program written on a PLC where the control program's execution speed is too slow.

Inclusive OR gate   A two-input logic device requiring both inputs to be in the same state to be satisfied before activating an output.

Input instruction   Located to the left of the output instructions on a rung and read data from the data table.

Instruction   Part of the ladder logic that can be interpreted by the processor to tell it what function to perform.

INT   Two-byte integer.

Interfacing   A process of connecting the processor or computer to external devices. This requires signal conditioning and the use of communication protocols.

Inverter   A logic device that changes what it sees on all inputs to the opposite at its output. 1s become 0s and 0s become 1s.

IP address   A specified IP address for every Ethernet device that is unique and is assigned by the manufacturer.

Isolated input module   A module that receives dry contacts as inputs, which the processor can recognize and change into two-state digital signals.

JIC   An acronym for Joint International Congress. The drafting symbols and layout for relay logic used in this text are written according to JIC standards.

Ladder-logic programming   A special language written to make it easy for people creating relay logic control to be able to program the PLC. Ladder-logic programming language uses symbols that graphically show what they are intended to do rather than using words. Most PLCs are programmed with this language.

Language module   A module that enables the user to program the processor using a language other than ladder logic. An example is BASIC.

Large PLC   A programmable logic controller with more than 1024 I/O terminals.

Logix5550   A model of ControlLogix controller.

Mask   Used to filter data in an instruction.

Medium PLC   A programmable logic controller with a maximum of 1024 I/O terminals.

Micro PLC   A programmable logic controller with a maximum of 32 I/O terminals.

Nano PLC   A programmable logic controller with a maximum of 16 I/O terminals.

NEMA standards   An acronym for National Electrical Manufacturers' Association standards.

Nested subroutine   A subroutine that calls another subroutine.

Nonretentive instruction   A ladder-logic output that doesn't hold its present value if the rung controlling it goes false. An example is a nonretentive timer, which resets to zero if the rung controlling it goes false.

Nonvolatile RAM   Random-access memory that will not be lost when the power is turned off. This is accomplished by battery backup, EEPROM, EPROM, or PROM.

Octal number   A number system with eight digits starting with 0 and ending with 7.

Optical isolation   A process in which a beam of light is used to electrically isolate the input from the output. The light is generated by an LED being turned off and on by the input. A phototransistor detects the light to turn the output off and on.

OR gate   A logic device that activates an output if any input is satisfied.

Output instruction   Located on the right-hand side of the rung and writing to the data table.

Overshoot   A process in which the set point is exceeded during automatic control. This means the system is not properly tuned.

PID   An acronym for proportional integral differential control, automatic control that uses three means simultaneously to determine the feedback. The feedback is determined by a value proportional to the error, plus how long the error is present, plus how fast the error is changing.

PID module   A module that enables the user to do proportional integral differential closed-loop automatic control. A set point and gains can be entered; if properly tuned, the module will hold the process at the desired set point.

PLC   An acronym that stands for programmable logic controller, which is a microprocessor-based device with built-in interfacing that can be programmed to automatically control a process. It is really a computer package for ease of interfacing that is designed to work in industrial environments. PLCs are often referred to as programmable controllers rather than programmable logic controllers, but the acronym PC is widely used to refer to personal computers. Adding the L differentiates the two and thus avoids confusion.

PLC module   A device that plugs into a PLC rack or chassis, which enables it to get power and communicate with the processor.

PLC processor   A computer designed specifically for programmable controllers. It supervises the action of the modules attached to it.

Process value   The input value to be kept at a set point by the automatic control; it is referred to as PV.

Program files   An area of memory where the ladder logic programs are stored. This is an orderly set of instructions for the processor to execute. It is in essence a file of binary-coded instructions.

Project files   An area of memory created in the ControlLogix controller that contains the programming and configuration information.

Protocol   A standard that defines the way a device will communicate with other devices. There are many protocols; examples are RS232 and DH485.

Rack   A piece of equipment used for determining addressing on PLC-5s. An I/O group is 32 bits, comprising 16 inputs and 16 outputs. A rack is 8 I/O groups.

REAL   Four-byte floating point.

Relay output module   A module that provides relay contacts that are isolated electrically to enable the user to use the contacts at any current and voltage within the contacts rating.

Reset   A term that means the same thing as integral in automatic control.

Retentive instruction   A ladder-logic output that does hold its present value if the rung controlling it goes false. An example is a retentive counter or timer that holds its accumulator value if the rung controlling it goes false.

Routine (ControlLogix)   A set of logic instructions in a single programming language, such as ladder logic. A routine is similar to the program files in an SLC 500 or a PLC-5.

RTD   An acronym for resistance temperature detector. It is used to measure temperature injunction with platinum, nickel, copper, and nickel-iron variable-resistance devices. Similar to the thermocouple module, it is designed to change the resistance read into scaled engineering units, such as degrees Celsius or Fahrenheit, and to work with PID instructions.

SCADA   An acronym for supervisory control and data acquisition.

Scaling   A process for changing raw binary data into engineering units humans can understand, or vice versa.

Scan   A cycle in which the processor reads the inputs, executes the program, and adjust outputs. This cycle is repeated over and over automatically by the processor.

Scan time   The time required to complete one cycle of a scan. A cycle in which the processor reads the inputs, executes the program, and adjust outputs.

Sequential control   A process that occurs in a sequential manner, where one operation must follow another. This is usually discrete control, which turns devices off and on as the process goes from beginning to end.

Set point   The value that the process value is to be held to by the automatic control function.

SFC   An acronym for sequential function charts, Allen-Bradley's name for its state diagrams, which are a technique for structuring a program for better documentation and troubleshooting.

SINT   One-byte integer.

Small PLC   A programmable logic controller with a maximum of 256 I/O terminals.

Starter   A contactor plus an overload device. It is used to start motors and to protect against overloads.

STI   An acronym for selectable time interrupt, a subroutine that executes on a time basis rather than an event basis.

Subroutines   Program files 3–255 in the SLC 500, which are scanned only when called upon by logic and can be used to break the program into smaller segments.

Synchronized axes module   A module that provides logic to synchronize multiple axes. Examples of applications are hydraulic tailgate loaders, forging machines, and roll-positioning operations. A mold-pressure module is used to connect with strain gauges for detecting mold cavity and hydraulic pressure. This information is used in a control program to improve repeatability.

Tags   Used by the Logix5550 controller to identify the memory location in the controller where data is stored.

Tasks   A means of scheduling and providing priority for how programs or a group of programs assigned to the task execute in a ControlLogix controller. There are two types of tasks, continuous and periodic. The Logix5550 controller supports up to 1 continuous and 32 periodic tasks. If there is no continuous task, there can be

32 periodic tasks, but if there is a continuous task, then there can only be 31 periodic tasks.

TCP/IP   An acronym for transmission control protocol/Internet protocol, a protocol for Ethernet.

Thermocouple module   Modules that enable you to read millivolt signals that standard analog modules cannot read. It interfaces with type J, K, E, R, S, B, C, and D thermocouples, but not all models have all these types. You need to check to see if the thermocouple you are using is covered by the model you buy. They are designed to change the millivolt signal into scaled engineering units such as degrees Celsius or degrees Fahrenheit and to work with PID instructions.

Thumb-wheel module   A module that reads a TTL BCD thumb-wheel device in parallel and changes it to uncoded binary for the processor to read.

Transducer   A device that produces either voltage or current proportional to some engineering units such as temperature (°C or °F), pressure ($lb/in.^2$), distance (cm), etc.

TTL input module   A module that receives two state inputs generated by TTL devices. This is 5-V logic.

TTL output module   A module that outputs two-state outputs to TTL devices. This is 5-V logic.

Tune   Using an automatic control system in which the feedback is adjusted to go quickly to the set point without overshoot.

Word   A continuous group of 16 bits forming a word for PLCs.

Index